INSTITUTE OF GEOLOGICAL SCIENCES
Natural Environment Research Council

British Regional Geology

Eastern England from the Tees to The Wash

SECOND EDITION

By Sir Peter Kent, DSc, PhD, FRS
with contributions by
G. D. Gaunt, BSc, PhD and C. J. Wood, BSc

Based on the First Edition by
Vernon Wilson, MSc, PhD, DIC

LONDON HER MAJESTY'S STATIONERY OFFICE 1980

The Institute of Geological Sciences was formed by the incorporation of the Geological Survey of Great Britain and the Geological Museum with Overseas Geological Surveys and is a constituent body of the Natural Environment Research Council

First published 1948
Second edition 1980

ISBN 0 11 884121 1

Foreword to Second Edition

The First Edition of this regional guide, under the title *East Yorkshire and Lincolnshire*, was written by the late Dr Vernon Wilson. It was published in 1948 and ran to ten impressions. Parts of the original account remain valid and are incorporated in the Second Edition, prepared by Sir Peter Kent while Chairman of the Natural Environment Research Council. I am much indebted to him for undertaking this task.

Since the First Edition was published much geological research has been carried out, many deep boreholes have been drilled and exploration for hydrocarbons in the adjacent North Sea has thrown new light on the region as a marginal part of an extensive basin of deposition. Account has been taken of this new information throughout the Second Edition.

Chapter 9 on the Upper Cretaceous has been written by Mr C. J. Wood and Chapter 12 on the Quaternary by Dr G. D. Gaunt. The latter has also contributed to Chapter 13 on Economic minerals and water supply. Sir Peter also wishes to acknowledge extensive help received from other officers of the Institute in the later stages of compilation. Dr C. Hill (British Museum, Natural History) contributed the section on Jurassic plants.

Our thanks are due to various companies concerned in the search for oil and gas on land and in the North Sea, and in particular to the British Petroleum Company, for the supply of information, some of it previously unpublished. Thanks are also extended to Dr J. A. Catt, Professor T. M. Harris, Miss M. W. Kendal, Dr L. F. Penny and Mr J. K. Wright, and to the following publishers, for permission to reproduce and draw upon illustrative material: Academic Press Inc. (London) Ltd [for Figure 13.7 from Annals of Botany (1943)]; British Museum (Natural History) [for Figures 5.1, 3–6; 7.1, 2, 4, and 13.8, 11, from British Mesozoic Fossils (1975, 5th edition), and for Figure 13.1–4, 9, 10, 12, from The Yorkshire Jurassic Flora]; Cambridge University Press [for Figure 12 taken from fig. 1 in Black (1928); the Scarborough Field Naturalists' Society [for Figure 8 taken from sections in Sylvester-Bradley (1953, pp. 22–3)]; Taylor and Francis, Ltd [for Figure 13.5, 6, from Annals and Magazine of Natural History (1952)]; and the Yorkshire Geological Society [for Figure 11 taken from fig. 1 in Black (1934), Figure 28 taken from fig. 59 in Penny (1974) and based on pl. 24 in Catt and Penny (1966), and Figure 16 based on pl. 12 and figs. 1–4 in Wright (1972)]. Mr Wright has also kindly provided new information which has been incorporated into Figure 16.

Institute of Geological Sciences
Exhibition Road
South Kensington
London SW7 2DE
3 October 1980

G. M. Brown
Director

An EXHIBIT illustrating the geology and scenery of the region described in this guide is set out in the Geological Museum, South Kensington, London.

Contents

Illustrations

[1] Numbers preceded by L refer to photographs in the IGS collections

1. Introduction

The region between the Tees and The Wash has an arbitrary boundary running down the middle of the Vale of York and the Trent Valley through or near Northallerton, Boroughbridge, York, Goole, Gainsborough and Newark to Grantham, whence it swings eastwards to Boston. It thus includes parts of the new counties of Cleveland and North Yorkshire, virtually the whole of Humberside and most of Lincolnshire. In this account, however, the Humber is taken as the boundary between Yorkshire and Lincolnshire, using these names in their older and more familiar senses.

When the previous edition of this 'Guide' was written, the region could be regarded as a distinct Mesozoic section of Britain, linked to the Midlands in the south by continuous outcrops, but not closely related to the older rocks of the Pennine region to the west. From a generation of deep boring activity in the search for coal and hydrocarbons, we now know a good deal about the continuation of the rocks of the Pennines at depth beneath the region and the relationship between the Mesozoic and Palaeozoic. Furthermore, the region is now also seen as the extremity of the very much larger North Sea Basin.

The information on the Palaeozoic rocks on land is inevitably localised and incomplete, while that on the Permian, Mesozoic and Tertiary rocks of the North Sea is much less detailed (being derived mostly from borehole cuttings and electric logs) than can be obtained from careful and deliberate study of exposures at outcrop. The surface and deep sources of information are thus complementary: the standard successions have been provided by the outcrops but the broader regional picture is constructed from deep borehole information both on land and offshore.

The scientific aspects of the region have been the subject of many valuable contributions to knowledge by naturalists, geologists and other men of science for many years, contributions so extensive that it is impossible here to give a complete survey. It will suffice to draw attention to a few of the outstanding advances made by workers who are no longer living. In the early days of geological science the Yorkshire coast was a happy hunting ground for the fossil collector and the more serious student. William Smith produced the first geological maps of Lincolnshire and east Yorkshire in the early part of the last century. The name of John Phillips is associated with early accounts of the rocks and fossils of the Yorkshire coast, and Tate and Blake's volume on the Yorkshire Lias is the starting point for all work on that subject. The Blea Wyke Beds and the Dogger have received special attention over a number of years from Rastall. Leckenby was a pioneer in investigating the sandstones, shales and oolites of the Scarborough district, while Blake and Hudleston are similarly linked with Corallian stratigraphy and palaeontology. Valuable contributions to our knowledge of the Upper Jurassic clays were made by Blake and Roberts, and Lamplugh's writings on the Speeton Clay are fundamental. Judd, Hill and Strahan laid the foundations of our knowledge of the intricate stratigraphy of the Lower Cretaceous deposits, and Jukes-Browne's account of the Chalk is a classic. Our knowledge of the Chalk was greatly enhanced when Rowe was able to extend his scheme of palaeontological zones for these rocks from the southern counties to Lincolnshire and Yorkshire. Jukes-Browne also took a prominent part in elucidating glacial problems, but the labours of Lamplugh, Clement Reid, Stather and Kendall will always remain the foundations of future work on glacial problems in Yorkshire.

The revision of this regional guide is made necessary by continuing research and changes in stratigraphic nomenclature in line with modern practice. The Yorkshire Geological Society's volume *The Geology and Mineral Resources of Yorkshire* (Rayner and Hemingway, 1974) is an outstanding compendium of data for the part of the region north of the Humber, and the smaller *Geology of Lincolnshire* (Swinnerton and Kent, 1976) of the Lincolnshire Naturalists' Union summarises data on the surface rocks farther south. Both have been drawn on in this account. Concepts of the regional structure, in particular of the Market Weighton Block which dominated Mesozoic deposition over much of the region, have changed considerably as a result of further investigation, and the trend of that structure is now more clearly defined although the fundamental cause of its movement remains obscure. Much more is known about the petrography of the rocks and about the conditions of deposition, but with this wealth of data a problem of space arises and additional sources are listed in the bibliography for those concerned with following up particular aspects.

Stratigraphic nomenclature is currently being regularised on an international basis. It is recognised that local rock formations often fail to fit standard time divisions, or to have exact equivalents elsewhere. Older names are thus often inappropriate, but the following account is designed to be used with existing Geological Survey maps and in consequence the traditional names are retained in this account side by side with their modern equivalents. Additionally the description refers to standard stages (time divisions) which are the fundamental basis for world-wide correlation.

Topography and regional geography

In general terms eastern England exhibits subdued topography, but in this region considerable diversity is found (Figure 1). Lincolnshire, despite its fenland, reaches more than 120 m OD in the Jurassic upland east of Grantham and more than 150 m in the Chalk Wolds, while the Jurassic hills of north-east Yorkshire rise to more than 450 m. The main physical features reflect the solid geology. In Lincolnshire the long scarps of eastward-dipping harder formations are separated by broad vales cut into the more easily eroded clays, while the more complex topography north-west of the Yorkshire Wolds is developed by dissection of faulted and gently folded rocks.

The lowlands of the Trent Valley and the Vale of York The western part of the region is formed by a belt of low ground extending from the Tees Valley southwards almost to Grantham, over which the Trent and Ouse river systems meander before emptying their water into the Humber Estuary. The Trent Valley is over 15 km wide in places and its carpet of Quaternary deposits is mainly underlain by Mercia Mudstone (Keuper Marl) with a border of Liassic clays to the east. Along the eastern side of this valley small tributaries of the

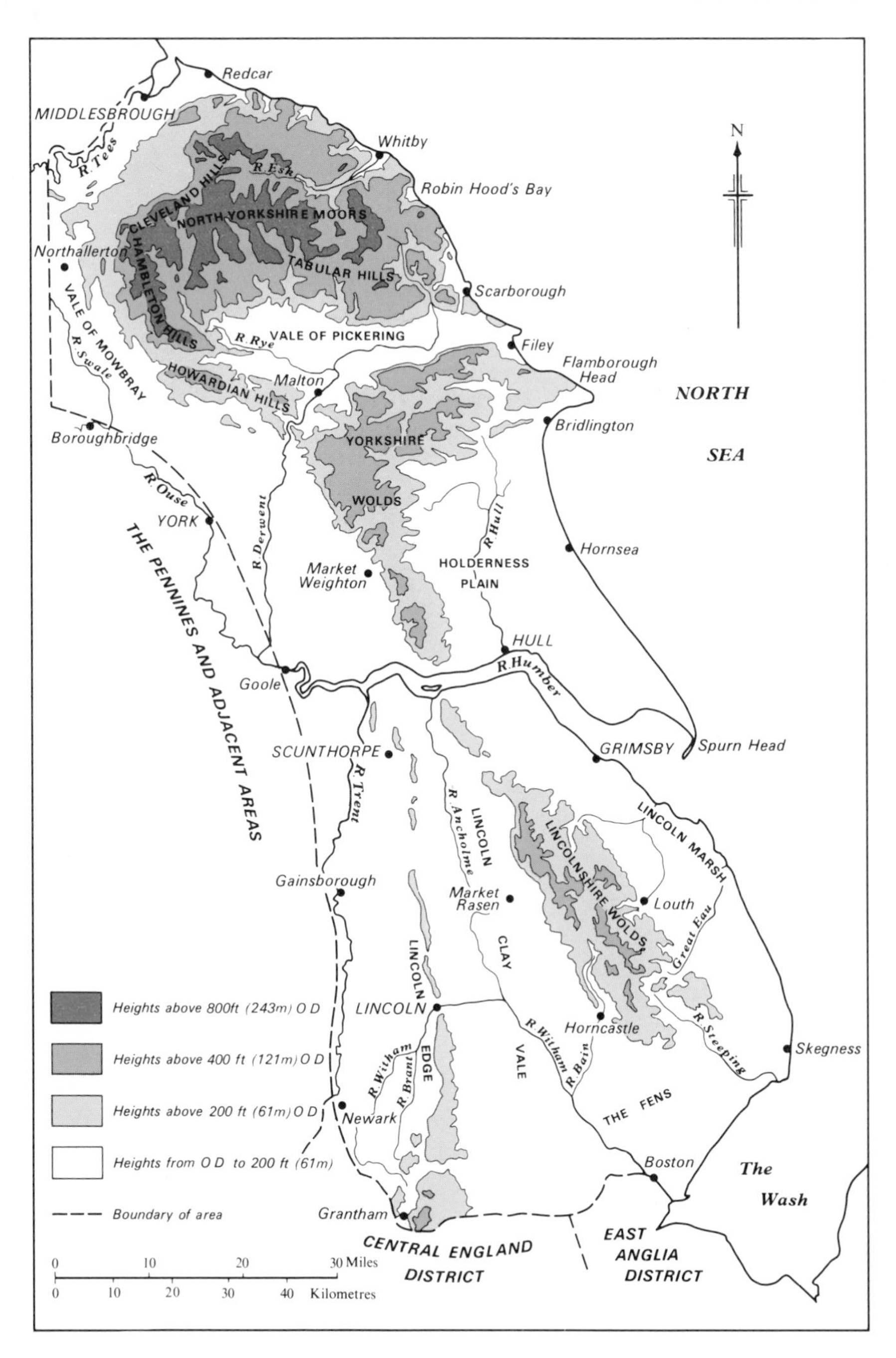

Figure 1 Physiographic map of the region

Rivers Trent and Witham have incised a series of minor valleys in a largely drift-free plain, so that the more resistant beds of the Lias, eg the Hydraulic Limestones and Granby Limestones, stand out and provide village sites which were distinctly drier than average in mediaeval times. The Vale of York is much wider than the Trent Valley and more heavily drift-covered so that the solid rocks exert no influence on the topography. Northwards from the Humber it narrows gradually to the vicinity of Northallerton, where the Jurassic scarp approaches the Pennines. This constricted northern part of the Vale of York is known as the Vale of Mowbray. Here low morainic hills form a divide between the drainage southwards to the Ouse and northwards to the Tees. Near York two prominent morainic ridges cross the Vale of York from the foot of the Wolds to the Pennines.

The lowlands were liable to frequent flooding down to historic times and consequently there is an extensive veneer of fertile alluvium. Much of this low-lying area has historically been rich pasture country, but in recent decades there has been a major change towards arable farming. Spreads of blown sand between Gainsborough and the Humber have been extensively afforested.

Lincoln Edge The broad floor of the Trent Valley is overlooked to the east by a prominent north–south scarp known as Lincoln Edge (Figure 2), which extends from Grantham to the Humber and is breached only by the Ancaster Gap and farther north by the Lincoln Gap. The lower slope of the scarp is formed by the Upper Lias, with the Northampton Sand and the Grantham Formation (Lower Estuarine Series) above, and the Lincolnshire Limestone at the top. Springs issuing from the permeable rocks above the Lias have determined the position of villages on the escarpment. On the dip slope the limestone is locally thin and inliers of the underlying sands and clays occur, eg around Nocton and Spital, bringing the water table to the surface, so that some of the small streams of the Lincoln Clay Vale rise within the limestone area. The bare plateau formed by the Lincolnshire Limestone is traversed by the Roman Ermine Street and has been farmed since Roman times (the present road plan is essentially a Roman layout). It remains largely arable, with emphasis on cereals and roots, grown in some of the largest fields in the country. Its dry, load-bearing character made it ideal for the development of wartime airfields.

Lincoln Clay Vale Between Lincoln Edge and the Lincolnshire Wolds is a lowland area known as the Lincoln Clay Vale or the Mid Clay Vale, extending from the Humber to the fenlands of south Lincolnshire. The northern part is drained by the River Ancholme and the southern part by the River Witham and its tributaries. The vale has resulted from the erosion of the soft Upper Jurassic clays and is heavily mantled by Quaternary deposits which largely conceal the solid rocks. The area is one of mixed agriculture, with forestry development on sandy drift-covered country near Market Rasen.

Lincolnshire and Yorkshire Wolds East of the Lincoln Clay Vale the ground rises to the Lincolnshire Wolds, a belt of dissected Chalk upland up to 15 km wide and about 70 km long, trending north-west. The Chalk escarpment is best developed in the central area, where it is fretted by streams; in the south it

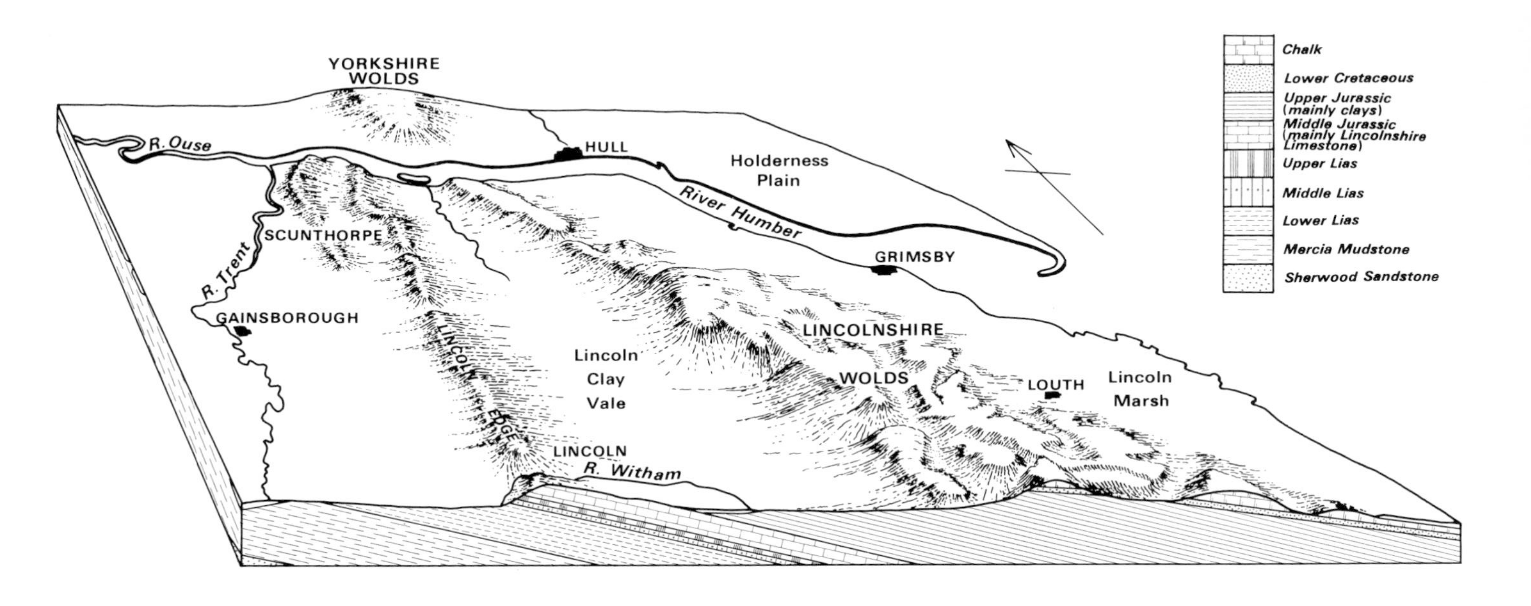

Figure 2 Block diagram of the physiography and solid geology of central and northern Lincolnshire

gradually ceases to be a prominent feature and its drainage is less mature, while in north Lincolnshire it is lower and more regular and has few streams.

Across the Humber the Chalk upland continues as the Yorkshire Wolds, which extend in a crescentic curve and end in the bold cliffs at Flamborough Head (see Plate 19). On the north the Wolds slope fairly steeply down to the broad Vale of Pickering, while on the west a narrow belt of Jurassic rocks prolongs their slope down to the Vale of York. Deep valleys carrying small streams trench the dip slopes and western escarpment. To the south-east the Yorkshire Wolds decline gradually to the gently undulating Holderness Plain.

Like Lincoln Edge, both the Lincolnshire and Yorkshire Wolds were the scene of much activity in Roman times, carrying major roads along their crests. The early sheep production has largely given place to arable farming. Also like Lincoln Edge they have provided sites for airfields.

Lincolnshire Fens The eastern part of the basin of the River Witham forms part of the Fens, which, limited by the 15-m contour and of an average elevation of 4.5 m OD, is a flat plain of recent marine and fluvial sediments resting on glacial deposits and occupying part of a shallow depression in the underlying Cretaceous and Jurassic rocks. The Fens have been changed from a vast morass to one of Britain's richest agricultural areas by drainage operations over many centuries. The Romans began the operation with a peripheral canal on the western side, the Car Dyke, to cut off the inflow of small streams and rivers; over the centuries the present system of drains and pumping stations has been developed.

Lincoln Marsh and the Holderness Plain The fenlands of south-east Lincolnshire pass north-eastwards into a flat marshy region up to 15 km wide, extending along the coast to the Humber. This is underlain by marine silts which rest upon glacial deposits. The Holderness Plain, the Yorkshire equivalent of Lincoln Marsh across the Humber, has low irregular relief about 3 to 10 m OD and terminates seawards in a long low line of cliffs averaging 10 m in height; its Chalk floor is buried beneath boulder clays, glacial sands and gravels and recent deposits. Marshes and meres were once common on this plain, but they have long since given place to cultivation, and only Hornsea Mere remains. The destruction of the Holderness cliffs by the sea goes on at disturbing speed, so that there is a steady loss of land and houses, and roads end on the cliff edges. Spurn Head is a peninsula built up by long-shore drift of material derived from erosion of these cliffs. It is unstable, and in historic times has been repeatedly broken through by the sea and subsequently reconstructed, changes documented by contemporaneous maps over the last 600 years.

The remaining part of the region, north and east of the Vale of York, presents considerable variation, and three main topographic areas can be recognised.

Howardian Hills The lowland to the north-east of York rises within 20 km to the Howardian Hills, a belt of irregular ridges about 6 km wide and up to 172 m OD in height. Their undulating topography has been largely determined by an intricate series of mainly east–west faults. The Howardian Hills and the Hambleton Hills to the north provide an area of small farms and mixed agriculture and woodland, which has changed relatively little in recent years. The limestone slopes bordering the Vale of Pickering yield good grain crops.

Vale of Pickering This broad plain lies between the dip slope of the Corallian rocks to the north and the Chalk escarpment to the south (Figure 3). It is a flat-floored west to east valley covered by a variable thickness of drift deposits, largely fluvial and lacustrine clays, which rest on Upper Jurassic mudstones. At its eastern end the Vale is cut off from the North Sea by a morainic barrier, and in the west it is linked with the Vale of York by the narrow Coxwold–Gilling Gap—a small trough-faulted valley separating the Howardian Hills from the Hambleton Hills. During the last (Devensian) glaciation meltwaters gathered in the Vale and formed an extensive lake which eventually topped the lowest point of the Howardian Hills near Malton and cut the Kirkham Gorge leading to the Vale of York. The flat floor of the Vale is conducive to large-scale agriculture, but the area lacks the fertility of the geologically comparable fenland farther south.

Cleveland Hills and North Yorkshire Moors The dissected plateau between the Vale of Pickering and the north Yorkshire coast includes the Cleveland Hills and North Yorkshire Moors. The topographic variations in this area reflect the nature of the rocks which crop out over its surface, much of which is above 300 m OD, though none of the summits exceeds 460 m. The greater part of the plateau, particularly in the north, consists of dissected hill ranges and wide moorlands bounded to the north and north-west by a steep scarp. It is an area mainly of Middle Jurassic sandstones and shales, with Liassic clays forming the lower slopes. Soils are poor; much of the area is devoted to grouse moors or sheep farming, but there is an extensive development of forestry on the lower slopes.

To the south this monotonous moorland slopes down to the foot of a well-defined broken escarpment, extending from the coast near Scarborough westwards to the Vale of Mowbray and marking the northern limit of the high, flat country known as the Tabular Hills. The general surface of these hills declines gradually southwards to the Vale of Pickering and is deeply incised by a number of gorges which have been cut by consequent streams flowing southward from the moorlands farther north. The Hambleton Hills, forming the western part of this plateau, terminate in a magnificent escarpment overlooking the Vale of Mowbray.

Drainage More than half of the region is drained by rivers that flow into the Humber.

A few small streams run down the scarp slopes of the Cleveland Hills into the River Tees, the River Esk takes a considerable amount of the surface water of the North Yorkshire Moors eastwards to the sea at Whitby, and the remainder is carried southwards by the Derwent. The small streams draining the North Yorkshire Moors above Hackness form the source of the present River Derwent, but prior to the last glaciation they flowed out to sea near Scalby Ness, north of Scarborough. In these times also the drainage from the areas round the Vale of Pickering gathered into a river whose course ran eastwards along the Vale to the sea near Filey. The blocking of these coastal exits

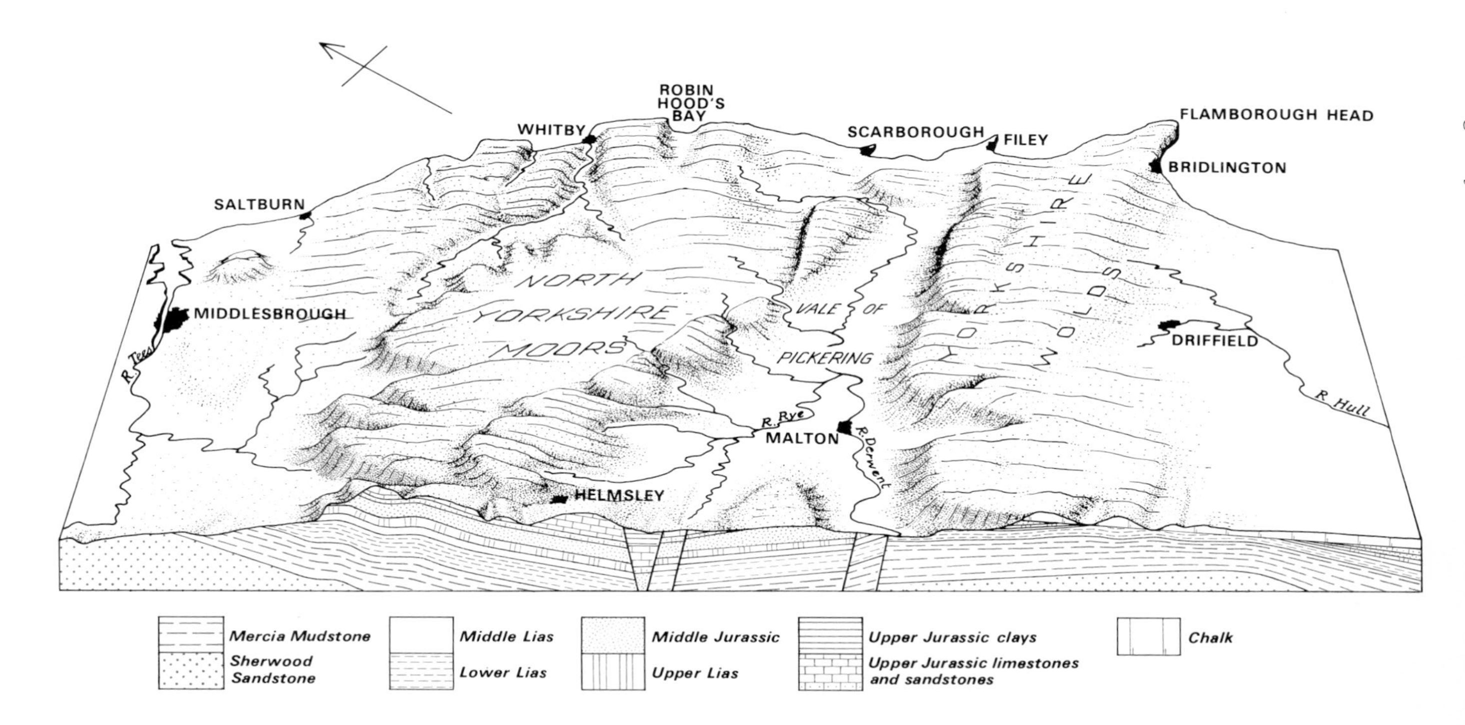

Figure 3 Block diagram of the physiography and solid geology of north-eastern Yorkshire

by ice and boulder clay ponded back these streams and eventually the accumulated water at Hackness escaped to the south through the present Forge Valley into Lake Pickering which overflowed through the Kirkham Gorge into the Vale of York.

In south-east Yorkshire the River Hull, fed by numerous small streams, carries the drainage from the Wolds southwards across the Holderness Plain to the Humber.

In Lincolnshire the River Trent flows northwards along the western edge of the region, but most of its water is drawn from the regions farther south and south-west. The only other river in Lincolnshire flowing into the Humber is the Ancholme, which drains the northern part of the Lincoln Clay Vale.

The River Witham flows northwards from Grantham at the foot of Lincoln Edge, but it passes through the Lincoln Gap into the Clay Vale and then follows a south-easterly course to The Wash. Numerous other small streams run off the Lincolnshire Wolds eastwards across the Lincoln Marsh to the coast.

The coast The coastline from Flamborough Head southwards is made up of simple and regular curves, the result of coastal erosion and deposition; the remaining third, bordering north-east Yorkshire, is rugged and irregular. About the Tees Estuary the coast is low and flat, but south-east of Saltburn it changes rapidly to high irregular cliffs, cleft at intervals by narrow defiles and small valleys. These cliffs, which continue to Flamborough Head, display magnificent sections of 'solid' rocks and their interest to geologists never wanes. At Saltburn, Liassic shales and sandstones form dark, unstable cliffs, but south-eastwards the harder Middle Jurassic rocks appear. Boulby Head near Staithes attains a height of 203 m and is one of the highest cliffs on the English coast. Towards Scarborough the coast is indented by small 'wykes' and bays. The Chalk gives rise to the magnificent headland and precipitous cliffs of Flamborough Head. The sea-caves and isolated 'stacks', a feature of the cliffs hereabouts, are due largely to wave action along joints, assisted by solution of the Chalk. Between Flamborough Head and Spurn Head, the regularity of the coast is due to the uniformly soft nature of the glacial deposits, and the low cliffs there are being rapidly eroded by the sea. Much of the Lincolnshire coast is formed of accreting marine alluvium, but between Mablethorpe and Skegness low cliffs consisting of glacial and later deposits are suffering erosion.

Human settlement and communications Human geography is closely related to the regional geology. The Romans chose the dry scarp lands for their roads, in part following far older roads of the Iron and Bronze ages. The alternative means of communication in early times was by water, and the distribution of village names of Scandinavian origin shows the extent to which invaders used the rivers for access. Cities like Lincoln and York grew up where Roman roads crossed major waterways; other early military settlements controlling fords, such as Littleborough in Lincolnshire and Brough on the Humber, lost their importance later.

Villages are located on belts of what was the drier ground in a generally wooded and waterlogged region, or are aligned along the spring lines below the limestone outcrops—along the foot of Lincoln Edge, along the Chalk scarp and on the flanks of the moorland heights of Cleveland.

When railways were built to link the towns they chose the natural gaps in the scarps—that of the Derwent in Yorkshire, the broad Humber Gap, the Barnetby Gap in the Lincolnshire Wolds and gaps at Lincoln, Ancaster and Grantham in Lincoln Edge. Standard-gauge rail networks totalling scores of kilometres were built also in the course of developing the Jurassic ironstones around Cleveland and Scunthorpe; these served their purpose for two or three generations but have lately been largely removed. Movement of bulk minerals (sand and gravel, limestone and chalk) is now mainly by road. Modern road builders with powerful machinery are less dependent on the topography, but attention to the surface geology in designing the layout of roads (and, incidentally, pipeline routes) can result in major economies in cost and effort.

Geological history

In broad terms the outcropping rocks of the region (Plate 1) were deposited in the western part of the North Sea Basin, but deep borings enable us to extend the history back into earlier and different regimes. The sequence of strata known to be present in the region is summarised in Table 1.

The oldest rocks known in the region are the metamorphic rocks of the Foston Borehole in Lincolnshire, the northernmost proving of Precambrian rocks, here partly volcanic in origin.

Elsewhere at depth steeply dipping quartzites have been encountered. These are believed to be of Lower Palaeozoic, possibly Cambrian, age and are thought to be relics of sediments widely deposited in a basin lying between the line of the Caledonian mountains and a foreland block which extended from the Baltic into East Anglia.

There is then a long gap in the historical evidence. If deposition took place in this interval any rocks which accumulated were subsequently eroded. The next indication of the geological history is provided by conglomerates and breccias, possibly of Devonian age, which accumulated in hollows on an irregular land surface. These deposits may be very local—they are known only in the Nocton area near Lincoln and at Eakring, west of the region.

During Dinantian times the sea transgressed over the whole region, and thick limestones were deposited in its clear water. Deltaic conditions subsequently spread across the region, with deposition of Millstone Grit sandstones and shales, mostly under fresh-water conditions, but occasional flooding by the sea resulted in the formation of marine intercalations. The Coal Measures, deposited under broadly comparable conditions, were distinguished by long periods during which peat, later to become coal, accumulated. As far as we know both Upper and Lower Carboniferous rocks were deposited across the whole region, but the amount of subsidence, in the earlier period at least, was much greater in north-east Yorkshire than on the East Midlands Shelf (south and east of the Market Weighton area) with consequent thicker sediments.

Late in the Carboniferous deposition was interrupted. The expression in this region of the more southerly Hercynian Orogeny involved uplift, faulting associated with gentle folding, and widespread erosion. Hundreds of metres of sediments were removed, particularly in east Yorkshire, and the region was

Table 1 Rocks known to be present in the region

<table>
<tr><td>Quaternary</td><td colspan="2">Tills (boulder clays), gravels, sands, clays, peats and cave deposits</td></tr>
<tr><td>Tertiary</td><td colspan="2">Cleveland Dyke</td></tr>
<tr><td>Upper Cretaceous</td><td colspan="2">Chalk</td></tr>
<tr><td rowspan="9">Lower Cretaceous</td><td colspan="2">Red Chalk</td></tr>
<tr><td colspan="2">Carstone</td></tr>
<tr><td>In Lincolnshire</td><td>In north Yorkshire</td></tr>
<tr><td>Sutterby Marl</td><td rowspan="6">Speeton Clay</td></tr>
<tr><td>Skegness Clay</td></tr>
<tr><td>Fulletby Beds (including Roach Stone)</td></tr>
<tr><td>Tealby Formation</td></tr>
<tr><td>Claxby Ironstone</td></tr>
<tr><td>Spilsby Sandstone (in part)</td></tr>
<tr><td rowspan="10">Upper Jurassic</td><td>Spilsby Sandstone (in part)</td><td></td></tr>
<tr><td>Kimmeridge Clay (including Elsham Sandstone)</td><td>Kimmeridge Clay</td></tr>
<tr><td rowspan="2">Ampthill Clay</td><td>Ampthill Clay</td></tr>
<tr><td>Corallian</td></tr>
<tr><td rowspan="3">Oxford Clay</td><td>Oxford Clay</td></tr>
<tr><td>Hackness Rock</td></tr>
<tr><td>Langdale Beds</td></tr>
<tr><td colspan="2">Kellaways Beds</td></tr>
<tr><td colspan="2">Upper Cornbrash</td></tr>
<tr><td colspan="2"></td></tr>
<tr><td rowspan="9">Middle Jurassic</td><td>Lower Cornbrash</td><td></td></tr>
<tr><td>Blisworth Clay</td><td></td></tr>
<tr><td>Great Oolite Limestone</td><td></td></tr>
<tr><td>Upper Estuarine Beds</td><td>Scalby Formation (formerly Upper Deltaic Series)</td></tr>
<tr><td></td><td>Scarborough Formation</td></tr>
<tr><td>Lincolnshire Limestone (Cave Oolite and subjacent strata in south Yorkshire)</td><td>Cloughton Formation (formerly Middle Deltaic Series, and including Yons Nab Beds, Whitwell Oolite and Millepore Bed)</td></tr>
<tr><td></td><td>Eller Beck Formation</td></tr>
<tr><td>Grantham Formation (formerly Lower Estuarine Series)</td><td>Saltwick Formation (formerly Lower Deltaic Series)</td></tr>
<tr><td>Northampton Sand</td><td>Dogger</td></tr>
<tr><td rowspan="3">Lower Jurassic</td><td colspan="2">Upper Lias</td></tr>
<tr><td colspan="2">Middle Lias (including Marlstone Rock in Lincolnshire and Cleveland Ironstone in north Yorkshire)</td></tr>
<tr><td colspan="2">Lower Lias (including Pecten Ironstone and Frodingham Ironstone in Lincolnshire)</td></tr>
</table>

Triassic	Penarth Group (*formerly Rhaetic*) Mercia Mudstone Group (*formerly Keuper Marl*) Sherwood Sandstone Group (*formerly Bunter Sandstone*)
Permian and older rocks proved only in boreholes	
Permian	Sherwood Sandstone Group (lowest part) Upper Marl (including evaporites in north and east Yorkshire) Upper Magnesian Limestone Middle Marl (including evaporites in north and east Yorkshire) Lower Magnesian Limestone Marl Slate and Lower Marl Basal Sands and Breccia
Carboniferous	Coal Measures Millstone Grit Carboniferous Limestone
Pre-Carboniferous	Conglomerates, quartzites and phyllites

reduced to a low, gently undulating plain. To the west it was flanked by hills, but to the east it extended with no great change across the area of the present North Sea.

Desert conditions followed, and early in Permian times dune sands were deposited by easterly winds, and torrent gravels accumulated in shallow wadis. This phase was probably fairly long lived, with slow erosion of remaining topographical irregularities. Eventually the Upper Permian Zechstein Sea flooded the region fairly rapidly and soon extended continuously from the Pennine flanks across the North Sea and much of northern Europe. Salinity fluctuated greatly in this sea; periods of fossiliferous dolomite deposition were separated by times of higher salinity when calcium sulphate and salts of sodium and potassium were deposited under 'evaporating dish' conditions.

The wide depositional basin established in the Permian continued to develop through the Mesozoic. During Triassic times the deposition of sands and gravels (the 'Bunter' facies) was followed by the accumulation of red clays (the 'Keuper' facies), for which modern analogues can be found only in continental desert basins.

Red bed deposition was ended when marine waters again flooded across the region in late Triassic times and the Penarth Beds (Rhaetic) were deposited. There was a short period of fluctuation before the long-lasting marine conditions of the Lower Jurassic were established. The region was then one of open seas, generally muddy, but occasionally clearing to allow thin shelly limestones to accumulate.

During the Middle Jurassic, sediments of deltaic type again spread across the region from the north, providing the shales and massive sandstones of the North Yorkshire Moors and the Cleveland Hills. South of the deltaic area were shallow, generally clear seas, although tongues of clastic sediment extended from the delta for variable distances into Lincolnshire. In these seas life abounded, and at times small patches of coral grew in clear water.

In the Upper Jurassic the region was entirely marine, and clay deposition was dominant. In middle to late Oxfordian times, however, the Cleveland Basin was the site of clear-water deposition with accumulation of the Corallian—mainly bedded limestones, but with coral reefs and sandstones (the 'calcareous grits'). Clay deposition was resumed later, but a thin lens of Kimmeridgian sand in north Lincolnshire suggests that shallows and perhaps a shoreline were not too distant towards the west. At the end of Jurassic times uplift, faulting and erosion took place.

The Lower Cretaceous rocks were deposited in a narrow belt linking southern England with boreal regions. Those of the Lincolnshire Wolds are shallow-water sandstones, limestones and ironstones with thin clays, whilst the corresponding Speeton Clay of Yorkshire shows sharp thickness changes indicative of control by faults which must predate the Chalk.

Widespread transgression followed. The Carstone sands were deposited in local hollows on the eroded surface, and the Red Chalk then spread over the region generally. The succeeding white Chalk is believed to have been deposited in clear seas which extended over most of Britain and lasted to the end of the Cretaceous Period.

Tectonic conditions did not remain uniform. The Cleveland Basin with its offshore continuation, the Sole Pit Trough, had been an area of heavy deposition during the Permian and much of the Mesozoic, but during Upper Cretaceous times became the site of broad differential uplift. Offshore this resulted in the thinning of the later parts of the Chalk; on land the evidence has been destroyed, but it is believed that at this time the Cleveland Anticline underwent its main uplift, rising nearly a thousand metres in relation to the previously buoyant area of Market Weighton to the south.

The Tertiary history can only be deduced from that of adjoining areas, for the landward part of this region appears to have been largely above sea level, perhaps slowly rising, and subject to erosion. Eocene rocks are present 50 km offshore, now forming an erosional scarp; originally they evidently extended farther west but probably never reached the line of the present coast. Offshore they were bevelled by Pliocene erosion and deposition, and erosion surfaces possibly of Pliocene age are recognised on land. In Quaternary times the landscape was modified by glacial activity and by extensive deposition from ice sheets and meltwaters. Later, the silts and peats of the Fens and Lincoln Marsh were deposited in sheltered bays. Silting up of the Lincoln Marsh was ended in medieval times when an offshore barrier was breached, but the Midlands rivers still bring their sediments to accumulate in the Fens and The Wash. Elsewhere—along most of the Yorkshire coast, and particularly in Holderness—coastal erosion is dominant and the resulting sediments are being deposited in the long peninsula of Spurn and in the wide spreads of sands off Grimsby and Immingham.

2. Concealed pre-Permian rocks (Precambrian to Carboniferous)

The rocks of the Triassic System are the oldest that crop out within the region, but boreholes have penetrated Palaeozoic and older rocks beneath for thicknesses of more than 1000 m.

The Trias is uniformly and conformably underlain by the Permian, which rests with a major unconformity on the eroded, largely planed-off surface of gently folded and faulted Carboniferous rocks, consisting in downward sequence of Coal Measures (Westphalian), Millstone Grit (Namurian) and Carboniferous Limestone (Dinantian). All three of these Carboniferous divisions are present south of the Humber, but the later part was commonly eroded off farther north. A few boreholes in the south have passed through the Carboniferous to reach Lower Palaeozoic and Precambrian rocks beneath.

These various rocks are briefly described in order, oldest first, in the following paragraphs.

Precambrian

Precambrian metamorphic rocks crop out in the Charnwood hills south-west of the region and continue eastwards at shallow depth along the northern edge of the Midlands Barrier to The Wash. Phyllitic rocks were reached in boreholes beneath the Millstone Grit at Sproxton, north-east of Melton Mowbray, and beneath Trias at Wittering and Glinton, near Peterborough, as well as in north Norfolk. Within the region phyllitic rocks compared with those of Charnwood were penetrated beneath Carboniferous Limestone at Foston, north-west of Grantham, on the crest of a plunging anticline extending north-westwards from the Charnwood–Wash ridge. The uplift of this ridge was evidently not only pre-Permian but also pre-Carboniferous, since it separates the subcrops of Lower Palaeozoic rocks to south and north.

Lower Palaeozoic

Fossiliferous Lower Palaeozoic rocks (Cambrian and Ordovician) are well known at depth south of the Midlands Barrier, but there is so far only one proving farther north, an occurrence of Lower Ordovician (Llanvirn/Arenig) in the offshore Burmah Well 47/29-1A east of The Wash. There are, however, several records of unfossiliferous quartzites beneath the Carboniferous in mid-Lincolnshire: these are thought to be Lower Palaeozoic, perhaps Cambrian (Kent, 1967). These beds are known as far north as Lincoln, but the corresponding stratigraphic levels have not been reached in north Lincolnshire and Yorkshire. In that area the depth to the aeromagnetic basement has been estimated at well over 6000 m, and, although such estimates are subject to major error, a possible implication is that pre-Carboniferous sediments are very thick under east Yorkshire.

Old Red Sandstone (?Devonian)

With the possible exception of beds below the Carboniferous Limestone in a borehole at Nocton, no Old Red Sandstone has been proved in the region. A little west of the Trent, however, the deep borehole Eakring 146 proved red conglomerates, interdigitating at the top with limestone of Lower Carboniferous $C_1/C_2/S_1$ age. These beds totalled 920 m in thickness. There is no evidence of date of the greater part of this succession, but a Devonian age for the lower part is a possibility. The conglomeratic beds rest unconformably on quartzitic sandstone broadly ascribed to the Lower Palaeozoic.

Carboniferous

Everywhere in the region the Permian is underlain by Carboniferous rocks (Table 2). The greater part of the subcrop is occupied by Coal Measures (continuing the Yorkshire and East Midlands Coalfield to the coast and beyond), but older Carboniferous rocks have been encountered immediately beneath the Permian across north Yorkshire from the Askrigg Block eastwards (Figure 4). Details there may be complex, since the area includes the Craven Fault Belt and its eastward continuation in the Howardian Hills, but the Permian is generally underlain by Millstone Grit, though locally by Carboniferous Limestone. Near the Tees there are again Coal Measures beneath the Permian.

Table 2 Major stratigraphical divisions of the Carboniferous

Chronostratigraphic divisions (Subsystem)	(Series)	Lithostratigraphic divisions
Silesian (Upper Carboniferous)	Stephanian	*not known in region*
	Westphalian	Coal Measures
	Namurian	Millstone Grit
Dinantian (Lower Carboniferous)	Viséan	Carboniferous Limestone
	Tournaisian	

Lower Carboniferous In northern England Lower Carboniferous deposition was characterised by strong differential epeirogenic subsidence, on a scale which exceeded any displacement due to later orogeny. This produced the well-known block-and-trough development of the Pennines, with the buoyant Alston, Askrigg and Derbyshire blocks generally the sites of relatively thin carbonate deposition, separated by rapidly subsiding basins or 'gulfs' which were the loci for much thicker clastic deposits.

The East Midlands Shelf beneath Lincolnshire was clearly an area of shallow deposition in the Lower Carboniferous, and by analogy this condition is presumed to continue northwards across the Market Weighton area to the Craven–Flamborough fault belt.

Farther north deep Carboniferous penetrations are infrequent and the situation is less clear. The Askrigg Block, however, ends eastwards between the Harlesey and Cleveland Hills borings, probably with a north–south margin running approximately from Stockton to Thirsk, and with a basinal development to the east coinciding broadly with the folded Cleveland area.

The Carboniferous development in the Eskdale–Lockton area is not adequately documented, but an expanded Millstone Grit sequence proved in the North Sea east of Scarborough suggests that the basinal facies of Cleveland continued eastwards in a belt marginal to the East Midlands Shelf; this however is still largely speculative.

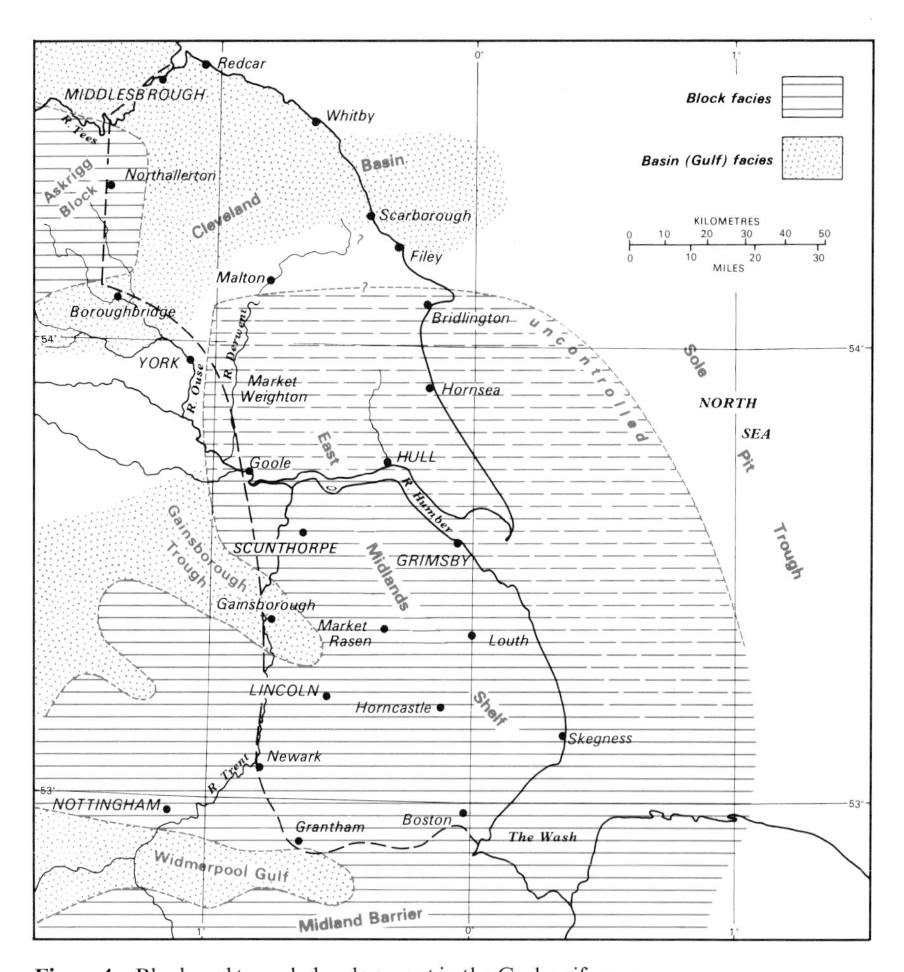

Figure 4 Block and trough development in the Carboniferous

Upper Carboniferous The Upper Carboniferous of northern England was deposited in the 'Pennine Delta', centred west of the Pennines and consisting of a series of lenses of sandstones, shales and coals which thins to the west, south-west and south-east. This delta was occasionally flooded by marine incursions, fairly numerous in the Namurian, less frequent in the Westphalian.

The easterly peripheral parts of the Pennine Delta lie within the Tees–Wash region. The main feature in the south-east (Howitt and Brunstrom, 1966) is strong attenuation of the non-marine elements, with persistence of thin marine bands, usually containing little more than the horny brachiopod *Lingula*. Progressive eastward thinning of the whole Carboniferous is also traceable into mid-Lincolnshire, so that the entire sequence (Coal Measures, Millstone Grit and Carboniferous Limestone) measures only 390 m at Stixwould, near Bardney. Of this, 297 m are Coal Measures.

The Carboniferous strata are, however, still present as far east as the coast, and are in fact known to thicken in east Lincolnshire (BP Well Tetney Lock No. 1), and the exploration of the North Sea has shown that the Upper Carboniferous at least is continuous from north-eastern England to the Low Countries and Germany. Relatively little is known of thickness variation in the North Sea, since the depth of penetration of the offshore boreholes beneath the Permian is usually minimal. The dating of the beds subcropping beneath the Permian indicates a broadly simple structural basin, with later beds—an upper red-coloured group which may be late Westphalian—limited to the central parts (Wills, 1973). There are minor perturbations in this regular arrangement offshore: in particular, Lower Palaeozoic rocks directly underlie the Permian north of the Dogger Bank (Amoseas Well 38/29-1) and this might be indicative of the position of a thinner belt between major delta lenses.

Igneous rocks

Carboniferous volcanic rocks (lavas, tuffs and agglomerates) occur on the south-western margin of the region and beyond, the thickest development, in excess of 100 m, being in the Lower Coal Measures.

Dolerite sills up to 30 m thick occur in both the Millstone Grit and Coal Measures in the neighbourhood of the Trent Valley southwards from Gainsborough. Some of these may be directly associated with the volcanics; others may be related to tensional flexing near the western edge of the East Midlands Shelf.

3. Permian

The Permian marked the beginning of a new regime in the region, the establishment of the Southern North Sea Basin, of which only the marginal part lies within the present land area. From Lower Permian times onwards through the Mesozoic this basin was continuous from the Pennine flank eastwards into mainland Europe, and from the East Anglian sector of the Wales–Brabant massif northwards. The regime was uninterrupted by any major period of folding, but depositional thicknesses were affected by epeirogenic movements—by differential subsidence relative to the basin margins and (especially in the Mesozoic) to the East Midlands Shelf and the Market Weighton area. Both thicknesses and the attitude of the rocks after deposition were modified by salt movements (halokinesis) originating in the Upper Permian Zechstein, which were characteristic farther east in the North Sea Basin, and also by a late Cretaceous inversion of the Cleveland and West Sole areas and by Tertiary movements. In contrast to these the Alpine Orogeny had much less marked effects in the present land area.

Permian rocks do not crop out in the region, and knowledge of them is therefore derived from boreholes.

Lower Permian

The period of erosion and planation which followed the Hercynian earth movements was protracted and thorough, and the surface of the Carboniferous rocks in the region was reduced to a peneplain. Lower Permian rocks deposited on this surface were non-marine.

None of the Lower Permian developments in the north-eastern England land area is of major thickness. The thicker sediments (10 to 30 m) are aeolian and fluviatile sands which were deposited in a very gentle depression centred over south Yorkshire. To the north of this, across north-east Yorkshire, was a low ridge (Smith, 1974) broadly corresponding with the contemporary outcrop of the Millstone Grit and Lower Carboniferous rocks, which were evidently more resistant to erosion than the softer Coal Measures to north and south. Over this ridge and in the immediately adjoining areas the Lower Permian is represented by thin breccias, partly discontinuous and usually less than a metre thick, made up of angular and subrounded fragments of Carboniferous rocks. Breccias in the south of the region have a considerable content of Precambrian rocks of Charnwood type, derived from the exposed edge of the Midlands Barrier.

The thicker Lower Permian sands (Rotliegendes) found offshore are important gas reservoirs: the West Sole and Rough gasfields are on the edge of this region, with the larger fields of the Leman–Indefatigable area in even thicker aeolian sands farther south-east.

Upper Permian (Zechstein)

In and around the North Sea Basin the Upper Permian began with a marine transgression, that of the Kupferschiefer and its English equivalent the Marl Slate, which is phenomenal for its wide extent in relation to its thickness (some ½ million square km in area but only 2 to 5 m thick). In much of Germany, the North Sea and eastern Yorkshire it is a dark bituminous dolomitic shale with fish remains and a modest marine fauna; farther west it is thicker, less bituminous and more calcareous. The German name Kupferschiefer relates to the widespread occurrence of metals in the shale; in England the metal content is not known to approach the economic level but is nevertheless abnormally high, and the stratum shows clearly on radioactivity logs of boreholes. Conditions of deposition of this distinctive bed have been much discussed (eg Smith, 1974) but it is essentially a marine deposit, laid down under stagnant conditions. The mortality of the fish may be related to poisoning by heavy phytoplankton production which also controlled the high bitumen content and the accumulation of metals.

The bulk of the outcropping Permian in north-eastern England is made up of dolomite and dolomitic limestone—the Magnesian Limestones—with associated anhydrite and salt becoming important down-dip. This is a highly variable group of rocks of which the correlation has defied resolution until recently, but Smith (1974) and Taylor and Colter (1975) have now related the outcrop development to the largely different facies down-dip on land and in the English sector of the Southern North Sea Basin.

The Upper Permian rocks were deposited in five widely recognisable major cycles. Each cycle began with deposition of carbonates (limestone and dolomite) followed by sulphates and chlorides, in some cases including potash minerals, and there is a lateral as well as an upward transition of the various evaporitic elements from the shoreline belt towards the centre of the basin.

The Lower Magnesian Limestone (Cycle Z1) is a basin-edge deposit—the shoreline lay only a little west of its present outcrop. It was essentially formed as a belt of oolite and pelleted carbonate sands, which gives place laterally to 150 m of anhydrite beneath east Yorkshire, and to thin limestones seaward of this.

In the second cycle (Z2) the depositional belts were situated farther seawards, but broadly similar relationships are recognised—beyond a landward development of red marl and anhydrite there is again a dolomite belt, around 100 m thick, which gives place basinwards to anhydrite and then to halite with the potash mineral polyhalite, which reaches a thickness in excess of 200 m. The second cycle is thus more complete than the first.

The three higher cycles (Z3 to Z5) are poorly developed at outcrop, but thicken gradually to some 200 m beneath the coast. In Z3 and Z4 the basinward development includes not only anhydrite and halite, but also potash minerals including sylvine and carnallite. It is these beds in Z3 which are now being developed by mining near Boulby on the Yorkshire coast as a potash source.

There are various other modifications of this complicated situation. Halokinesis is well developed beneath the North Sea from the West Sole Trough eastwards, where it produces a range of structures from salt pillows to piercement salt plugs. The latter may penetrate thousands of metres of overlying rocks and were moving in some cases from Middle Triassic times onwards, markedly affecting the depositional thicknesses of later sediments (Kent, 1967; Brunstrom and Walmsley, 1969). These piercement structures are best developed in the area of the Southern North Sea troughs and are not known on land, but the sharp variation in the thickness of the salt-bearing beds in east Yorkshire is thought to be due to lateral flowage which is detectable in the arrangements of the mineral grains.

4. Triassic

Rocks of the Triassic System crop out along the western edge of the region from Middlesbrough to Newark, but are largely hidden by alluvium and glacial deposits. Most of the detailed knowledge of the sequence is consequently derived from boreholes, which demonstrate also that the rocks continue eastwards under the later cover into the North Sea.

The Triassic rocks of north-western Europe were deposited during a major arid episode. At the beginning of Triassic times much of Britain was occupied by ranges of low hills, but the low ground of north-eastern England had been flooded by the Permian sea, which left behind extensive mud flats. As in modern Australia and the Middle East, torrent gravels, products of erosion of the high ground, accumulated round the edges of the basin, while sands and muds were swept into the lower ground where, from time to time, evaporites were also deposited.

This basin was occupied at least twice by the sea. On the first known occasion the Muschelkalk Sea of Germany extended brackish tongues into eastern England, giving rise to dolomitic intercalations in the red clay sequence of the Mercia Mudstone Group, and enabled the marine brachiopod *Lingula* to establish itself briefly in Nottinghamshire. The second important marine incursion was the terminal episode of the Trias, when the Rhaetian sea swept quickly across the Triassic basin, bringing with it a varied and somewhat specialised marine fauna.

The Trias in Britain has traditionally been divided into two parts—Bunter and Keuper; a third unit, the Rhaetic, has on occasion been included in the Jurassic but is now grouped with the Triassic. It has been found, however, that the traditional British units differ importantly in age and facies from European deposits with the same names. Bunter, Keuper and Rhaetic have therefore now ceased to be formal names in Britain and are replaced by the Sherwood Sandstone, Mercia Mudstone and Penarth groups—with adjusted boundaries which more accurately reflect gross lithology.

A full account of the Trias of this region has been compiled by Warrington (1974).

Sherwood Sandstone Group (former Bunter Sandstone)

Over east Yorkshire and Lincolnshire the marine evaporitic Permian is followed conformably and gradationally by the mainly red sandstones of the Sherwood Sandstone Group, 150 m thick in the south and over 350 m in the north. While the youngest Permian evaporites were being deposited in the north, red sandstones were already accumulating in the south, adding further to the problems of correlation and nomenclature.

The dominant feature of the English Midlands during the early Triassic was the tremendous flood of gravel washed into the basin from the south—the

Bunter Pebble Beds of the original terminology—which constitutes the upper part of the Sherwood Sandstone Group. The main source of this detritus was south of Britain, for the local hill masses made only small contributions of torrent gravel. In north Lincolnshire the Pebble Beds give place northwards to finer sands, for the coarse material was dropped first by the debouching rivers. In Yorkshire the whole mass of the Sherwood Sandstone Group is comprised of reddish medium to fine-grained sandstone, becoming progressively softer, muddier and less porous northwards where the rivers deposited their silt and clay load. In line with this change the group, which is an outstandingly important aquifer in the Midlands, loses a large part of its ability to hold and yield water in the muddier facies of Yorkshire.

Mercia Mudstone Group (principally argillaceous facies of the former Keuper)

It is not known precisely why the flood of coarse material from the south was cut off, but from the time of widespread minor earth movements which resulted in a minor unconformity affecting the North Sea Basin and adjoining areas (the Hardegsen disconformity of Germany), the deposition of argillaceous and silty rocks became dominant. Deposition of the Mercia Mudstone Group began earlier in the middle of the basin, where rocks of this facies were accumulating while arenaceous rocks were still being laid down on the margins of the basin. In these latter areas, greyish conglomeratic sandstones rest on an eroded surface of the Pebble Beds. This surface is notable farther west for the occurrence of facetted pebbles (dreikanter or ventifacts) sculptured by sand blasting during the pause in sedimentation caused by the minor earth movements.

The marginal facies of the Mercia Mudstone Group also includes the arenaceous part of the former 'Waterstones', now the Colwick Formation, an alternation of fine micaceous sandstones and bedded mudstones up to 60 m thick. These beds have many features, including salt pseudomorphs, characteristic of the argillaceous facies of the group, into which they pass laterally. The Colwick Formation is notable also for the discovery of shoals of fossil fish which died huddled together in pools as temporary lakes evaporated, and for the occurrence of *Lingula*, found in quantity at a single exposure near Eakring.

Although the main body of the Mercia Mudstone is red 'marl' (variably red calcareous silt and clay, with minor intercalations of green and greyish beds) there is some diversification. The red mudstone sequence of 180 to 320 m is interrupted by units of bedded sandstone and siltstone (skerries) which are particularly prominent towards the base in strata laterally equivalent to the Colwick Formation. The skerries all thin out across Lincolnshire and Yorkshire towards the centre of the basin. In this direction the lower part of the group becomes increasingly dolomitic (Balchin and Ridd, 1970), reflecting passage into the Muschelkalk facies.

Development of evaporites is a further basinward variation within the mudstones. The region lacks the thick salts of the 'Keuper Marl' of Cheshire (although their approximate equivalents are developed 50 to 80 km offshore beneath the North Sea) but an older salt unit some 50 m thick is present in the basal Mercia Mudstone in east Yorkshire. This was initially seen in Fordon No. 1 Borehole, near Bridlington, and subsequently found in other boreholes near the coast. It is correlated with the Rötsalinar of the German Bunter.

Additionally, gypsum and anhydrite are present, interbedded with the mudstone, especially at an horizon 24 to 36 m below the top of the group. Gypsum has been extensively worked at outcrop in south and east Nottinghamshire, between York and Malton and near Northallerton.

The topmost part of the Mercia Mudstone is the Tea Green Marl, up to 10 m of green and yellowish mudstone, variably gypsiferous and much resembling the normal red mudstone facies except in colour. This has been considered to be the chemically reduced top of the red beds, but it may alternatively be an original transitional facies to the marine beds above.

Rarity of fossils in the Mercia Mudstone may be partly related to the paucity of exposures, but borehole cores in the upper part (principally in the Tea Green Marl) have yielded scattered fish remains, precursors of the more normal marine conditions of the Penarth Group.

Penarth Group (former Rhaetic)

The change from red beds to the grey marine strata of the Penarth Group (Rhaetian) and the Jurassic is one of the most striking in the geological column. Clays continued to be the dominant sediment, as in the Mercia Mudstone, but they were deposited under strongly reducing conditions, in waters that were at times rich in marine life.

There are indications that the earliest positive movements of the Market Weighton Block, which was later to divide the Mesozoic basin into two subbasins, took place in Rhaetian times (Kent, 1955).

The first sediments deposited during the marine incursion were the dark grey to black shales of the Westbury Formation, some 2 to 11 m thick. They are generally finely laminated and highly fissile, with a variable development of thin sandstones, which are locally rich in fish scales and teeth and also contain reptilian debris—the 'Rhaetic Bone Beds'. The bone-bed sandstones lack the conglomeratic developments found in southern Britain, since most of the region was distant from shorelines. The beds have been variously ascribed to catastrophic extinction of the fauna due to salinity changes, or alternatively to slow deposition and condensation of the deposit. Probably both factors were involved; multiplication of bone-bed sandstones in the thicker developments argues against simple condensation, but the sandstones indicate phases of strong currents which swept in the coarser sediment, prevented deposition of clay, and may well have been associated with lowering of salinity causing abnormal mortality of the fish.

The shales contain a restricted molluscan fauna, mostly small forms and commonly with a predominance of one species on an individual bedding plane. The bivalve *Rhaetavicula* (formerly *Pteria*) *contorta* is the most characteristic. In Lincolnshire an impure limestone developed in the upper part of the shales contains a clear-water fauna including *Chlamys valoniensis.*

The marine Westbury Formation is overlain by the Cotham Member of the Lilstock Formation. This includes not only pale greyish green mudstone, as in the Midlands, but also chocolate-brown and reddish mudstone in northern Lincolnshire, east Yorkshire and beneath the North Sea, indicating a partial reversion to Mercia Mudstone conditions. The thickness of the Cotham Member varies from 2 to 15 m in this region, with the maximum in central Lincolnshire. The fauna is usually limited to *Euestheria minuta*, but in the uppermost part at Kettlethorpe in west Lincolnshire, large limestone nodules have yielded a marine fauna comparable to that in the Langport Member (formerly 'White Lias'), which overlies the Cotham Member in southern Britain. The typical bivalves *Modiolus langportensis* and *Dimyopsis intusstriatus* are among fossils present. Recent boreholes in the Humber and Acklam areas of Yorkshire have shown that the Cotham Member is there overlain by grey fissile mudstones of Lias lithology, which, near their base, contain bivalves of Rhaetian affinity, for example, *Gervillea praecursor*.

The Penarth Group as a whole expands progressively from a few metres near Grantham to nearly 20 m in west central Lincolnshire. It is 11.5 m in Cockle Pits Borehole north of the Humber and, except over the Market Weighton Block, where it may be less than 5 m, is 9 to 15 m where proved elsewhere in Yorkshire.

Eastwards beneath the North Sea the black shales of the Westbury Formation extend for 20 to 30 km, but beyond this become difficult to distinguish in borehole cuttings and the Penarth Group consists dominantly of greenish and brownish mudstones with a variable sandstone near the base. The thickness increases to 20 to 30 m. The poor development or absence of marine beds is in line with evidence from southern England that the tongue of marine waters which flooded the late Triassic basin came from the south, via the Dorset–Somerset area, and was petering out by the time it reached the southern North Sea.

5. Jurassic: Lower Jurassic

The Jurassic rocks crop out over the whole of the Cleveland Hills, Howardian Hills and North Yorkshire Moors, form a narrow belt along the western margin of the Yorkshire Wolds and continue with a broadening outcrop from Market Weighton southwards across Lincolnshire into the Midlands.

The contrast between the complex outcrop patterns north of Market Weighton and the nearly straight outcrops of southern Yorkshire and Lincolnshire reflects the geological structure. The former area has been subject to gentle folding from Jurassic times onwards, whereas the latter forms part of the East Midlands Shelf and has been gently tilted eastwards but has otherwise remained almost completely unfolded. Progressive thinning of the Jurassic rocks from the Fens northwards to Market Weighton, plus a gently increasing overlap by the Cretaceous, results in narrowing of the outcrops northwards across the shelf. Attenuation and overlap reach a maximum on the Market Weighton Block, which is in fact the northernmost element of the East Midlands Shelf abutting against the Cleveland Basin.

Geologically the region thus falls into two distinct provinces, north and south of the Market Weighton area, and the description follows this distinction.

As elsewhere in north-west Europe the British Jurassic contains a variety of lithologies deposited under dominantly marine conditions. The sequence shows an alternation of shales and clays with limestones and ironstones, siltstones and sandstones. In the Tees–Wash region there is a thick wedge of deltaic beds in the Middle Jurassic of the Cleveland Basin and there are other differences from the southern England succession, including a tendency to northern faunas and the absence of some characteristic southern fossils.

Traditionally the Lower Jurassic rocks have been divided into Lower, Middle and Upper Lias, but there is the complication that in Yorkshire the middle division originally spanned a wider range than elsewhere. The relationship of lithology to stages and zones is shown in Table 3.

Lower Lias

In very broad terms the Lower Lias is a clay succession with interbedded minor limestones. However, many authors from Tate and Blake (1876) to Bairstow (1969) have shown that the Yorkshire coast sequence is variable in detail and can be subdivided lithologically. Boreholes, temporary exposures and recent mapping have shown this to be true in Lincolnshire also.

The initial influx of the Jurassic sea following the mainly non-marine late Rhaetian episode appears to have been effectively simultaneous across the region. The earliest Hettangian beds, the 'Hydraulic Limestones', show an alternation of shales with shelly or argillaceous partly concretionary limestones and are 5 to 10 m thick; the basal part (the Pre-*planorbis* Beds) contains

Table 3 Zonal correlation of Lower Jurassic rocks of the region

Stage	Substage	Zone
Toarcian	Yeovilian	*Dumortieria levesquei*
		Grammoceras thouarsense
		Haugia variabilis
	Whitbian	*Hildoceras bifrons*
		Harpoceras falciferum
		Dactylioceras tenuicostatum
Pliensbachian	Domerian	*Pleuroceras spinatum*
		Amaltheus margaritatus
	Carixian	*Prodactylioceras davoei*
		Tragophylloceras ibex
		Uptonia jamesoni
Sinemurian	Upper Sinemurian	*Echioceras raricostatum*
		Oxynoticeras oxynotum
		Asteroceras obtusum
	Lower Sinemurian	*Caenisites turneri*
		Arnioceras semicostatum
		Arietites bucklandi
Hettangian		*Schlotheimia angulata*
		Alsatites liasicus
		Psiloceras planorbis

Lithostratigraphy		
North-east Yorkshire	South-east Yorkshire & north Lincolnshire	Central and south Lincolnshire
Blea Wyke Beds: Yellow Beds, Grey Beds	(Absent)	(Absent)
striatulus Shales		
Peak Shales		
Cement Shales Main Alum Shales Hard Shales		
Bituminous Shales	(Mudstones)	(Mudstones)
Jet Rock		
Grey Shales		'Transition Bed'
Cleveland Ironstone	Marlstone Rock	Marlstone Rock
Staithes Formation	(Mudstones)	(Mudstones with ironstone nodules)
Ironstone Shales	*Pecten* Ironstone	(Mudstones with thin siltstones)
Pyritous Shales	(Mudstones with thin siltstones)	
Siliceous Shales		Sand Rock (Mudstones)
Calcareous Shales	Frodingham Ironstone	Bassingham Calcareous Sandstone (Mudstones) (Mudstones with ferruginous limestones including 'Plungar Ironstone')
	(Mudstones with thin limestones)	(Mudstones) Granby Limestones (Mudstones)
	'Hydraulic Limestones'	'Hydraulic Limestones'

a marine fauna of foraminifera, ostracods, echinoids and bivalves. Ammonites arrived later, first the zonal index *Psiloceras planorbis*, then species of the subgenus *Caloceras*. Bivalves occur in quantity—particularly *Liostrea hisingeri*, *Modiolus minimus* and *Pteromya tatei*. The marine reptiles *Ichthyosaurus* (Plate 2) and *Plesiosaurus* have been found from time to time in working quarries in the 'Hydraulic Limestones' to the south-west of the region (Barrow-on-Soar and Barnstone (Plate 3.1)).

The 'Hydraulic Limestones' continue in strength into south Yorkshire, forming a strong feature near North Cave, but they are less distinctive farther north and on the Teesside coast are blanketed by Boulder Clay.

Over south and mid-Lincolnshire the phase of shallow-water limestone deposition was followed by deepening, with accumulation of 15 to 20 m of dark, pyritic shales which contain the zonal ammonites *Alsatites liasicus* and *Schlotheimia angulata* (Plate 7). Infrequent shelly bands consist largely of *Cardinia*. These shales thin in north Lincolnshire, with lateral passage into a clay and limestone alternation lithologically similar to the beds above, and there is apparently no separate clay interval at this level in Yorkshire.

Over most of Lincolnshire the upper part of the *angulata* Zone and the lower part of the *bucklandi* Zone are developed as shales alternating with soft argillaceous limestones, the 15-m Granby Limestones. *Gryphaea arcuata obliquata* and, at higher levels, *G. a. incurva* occur in great abundance, with common large *Plagiostoma giganteum* (Figure 5) and (in the middle part) the coral *Montlivaltia haimei*. Large specimens of *Schlotheimia* are common in the lower beds; *Coroniceras* and its allies occur above in the *bucklandi* Zone. A widespread feature in Lincolnshire is the occurrence of the truncated-conical shelly nodules known as *Kulindrichnus*. The upper part of the *bucklandi* Zone is represented in the south by clays with minor ferruginous beds.

Ferruginous strata develop in the basal Lias towards the Market Weighton Block. In north Lincolnshire limonite ooliths are abundant in beds equivalent to the Granby Limestones, and in Yorkshire (particularly around Givendale) limonite oolite occurs in the *angulata* Zone and possibly earlier. There is much ferruginous alteration in the detrital Lias limestones of the latter area, so that both primary and secondary deposition of iron were probably involved. This variant extends eastwards at depth to the coast.

In north Yorkshire the whole of the Hettangian and part of the Sinemurian sequence consist of shales with thin limestones (the Calcareous Shales) with much the same range of fossils as in Lincolnshire. The lower part is well exposed on the coast near Redcar, and the upper part, which extends up into the *obtusum* Zone, is exposed in Robin Hood's Bay, where it is succeeded by the Siliceous Shales of *obtusum* to *raricostatum* Zone age.

Lincolnshire shows two contrasting facies above the *bucklandi* Zone. In south Lincolnshire there are 30 to 40 m of clays with thick ferruginous limestones, of which one (the 'Plungar Ironstone' of older authors) is strongly limonitic locally. *Arnioceras semicostatum* and its allies, together with species of *Gryphaea*, occur abundantly; the zonal index *Caenisites turneri* has been found near Grantham, and *Eparietites* in the higher part indicates the *obtusum* Zone. In north Lincolnshire, near Scunthorpe, the clay and limestone sequence is succeeded by the Frodingham Ironstone, which extends from the upper part of the *semicostatum* Zone to the basal part of the *oxynotum* Zone.

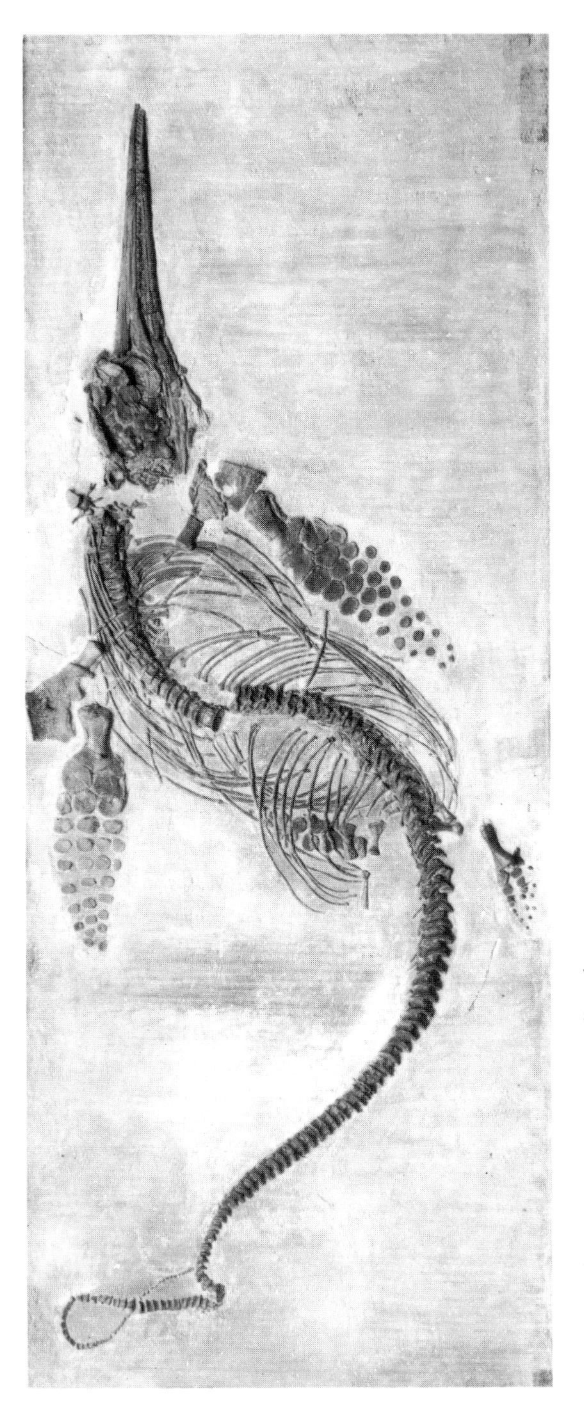

Plate 2 *Ichthyosaurus tenuirostris* Conybeare;
Lower Lias, This specimen is about 2.75 m long and is from Street, Somerset

The Frodingham Ironstone is a major economic deposit, the basis for the industrial complex of north Lincolnshire and the only one of the Jurassic ironstones in this region still exploited. It has been quarried over a distance of 12 km along the strike (see Plate 3.2) and has been mined for several kilometres to the east. It is up to 10 m thick and is lenticular with the area of maximum thickness extending from Scunthorpe eastwards below the Chalk outcrop towards Immingham. It is only thinly represented north of the Humber and south of Grimsby, so that it lies in a west–east belt, parallel to the trend of the Market Weighton Block. All but the lowermost part is normally worked as ore. It is essentially a chamositic and sideritic oolite, rich in shell material in most beds, with minor local intercalations of clay and limestone. At outcrop it weathers to a limonite oolite, but the weathered and therefore enriched section of the orefield was mined away in the early stages of development.

Although the incoming of iron-precipitating conditions was heralded by the occurrence of limonite ooliths in the underlying limestones in north Lincolnshire and south Yorkshire, the depositional conditions that produced ore-grade material began in the *semicostatum* Zone (*Pararnioceras alcinoe* and *Arnioceras* occur in the lowest bed). The *obtusum* and *turneri* zones are represented above, locally with rich ammonite faunas including common *Eparietites*. *Oxynoticeras* has been found in the upper metre of the ironstone, but the *oxynotum* Zone is mainly represented by the overlying silty clays. Other mollusca are usually abundant in the ironstone. *Gryphaea maccullochii* and *Cardinia* occur in quantity, especially in the middle beds; *Pholadomya*, limids and pectinids are found in lesser numbers. In contrast to the Middle Lias ironstones, brachiopods are infrequent.

The Frodingham Ironstone dips gently eastwards beneath an increasing thickness of overburden and the deeper opencast mines expose the whole of the succeeding Lower Lias, some 30 m thick. This part of the sequence consists of clays with silty layers, often rich in ammonites, with the *Pecten* Ironstone in the upper part.

Mid-Lincolnshire lacks ferruginous strata equivalent to the Frodingham Ironstone of the north and the Ferruginous Limestones of the south. Near Lincoln the highest bed of the latter passes into the Bassingham Calcareous Sandstone (approximately *obtusum* Zone age), the first of several influxes of sand into the basin. This bed, and a higher more persistent calcareous sandstone (the Sand Rock) close to the *oxynotum/raricostatum* zonal boundary which crops out near Brant Broughton and Broxholme, form minor cuesta features in the Lias plain.

In both mid-Lincolnshire and Yorkshire the early Pliensbachian (*jamesoni* Zone) was marked by deposition of fine silts, rich in muscovite, which weather much paler than the clays above and below. They contain reddish-weathering siderite nodules and are associated with a shallow-water or intertidal fauna including common large *Pinna*.

These intercalations are the result of minor influxes of silt and sand which spread widely across the basin at times of general shallowing, presumably from a land area to the west. Their repetition in the dominantly clay series suggests a broad rhythmic control of sedimentation. In north-east Yorkshire there was additionally a general increase in sandiness towards the end of

Figure 5 Lower and Middle Lias bivalves and brachiopods (natural size, except 1, 5 and 7)
1 *Pseudopecten equivalvis* (J. Sowerby), $\times \frac{1}{2}$; **2** *Oxytoma inequivalvis* (J. Sowerby); **3a–c** *Tetrarhynchia tetrahedra* (J. Sowerby); **4** *Protocardia truncata* (J. de C. Sowerby); **5** *Gryphaea arcuata* Lamarck, $\times \frac{3}{4}$; **6** *Spiriferina walcotti* (J. Sowerby); **7** *Plagiostoma giganteum* J. Sowerby, $\times \frac{3}{4}$; **8** *Oxytoma* (*Palmoxytoma*) *cygnipes* (Young & Bird).

Lower Lias deposition, well developed in the *davoei* Zone; this is not known in Lincolnshire, and was presumably due to an influx from a more northerly direction. The section of the Lias in the Cleveland Hills Borehole shows, however, that in the north-west most of the upper part of the Lower Lias is sandy, so that westerly sources were clearly significant.

In both Yorkshire and Lincolnshire ironstone nodules are common in the upper half of the Lower Lias, and where limestones occur these are usually composed of shell debris. The iron content of the higher beds reached a maximum with deposition of the *Pecten* Ironstone in north Lincolnshire and south Yorkshire; this comprises up to 2 m of sideritic ironstone crowded with valves of *Pseudopecten equivalvis* (Plate 4; Figure 5). Precise evidence of date is scanty but it appears to span the *jamesoni/ibex* zonal boundary.

In the upper part of the Lower Lias the faunas are often very rich, although in many cases confined to particular bands, usually limestones or ironstones. This is in part a preservation factor; in the clays they may have been decalcified during diagenesis and subsequently weathered away. Survival of microfaunas in otherwise barren beds, however, shows that distribution-failure is a factor also. Ammonites occur throughout; in Lincolnshire they are particularly abundant and varied in the *ibex* Zone, and well preserved iridescent 'capricorn' ammonites *Aegoceras* (*Oistoceras*) and *A.* (*Androgynoceras*) (Plate 7) occur in quantity in the *davoei* Zone. The zonal ammonite *Prodactylioceras davoei* is rarer in Lincolnshire than in southern England and is not known in Yorkshire. Among the gastropods, small ceritheid genera occur in abundance in Lincolnshire in the Pliensbachian Lower Lias. From the *jamesoni* Zone upwards belemnites are often the most prominent fossils, but their taxonomy and sequence has not been worked out.

Middle Lias

The Middle Lias, largely assigned to the Domerian Substage of the Pliensbachian, is bipartite, with an ironstone named the Marlstone Rock (in Lincolnshire) and the Cleveland Ironstone (in Yorkshire) above and sands or clays below. The latter beds are represented in north-east Yorkshire by the Staithes Formation.

This formation, comprising up to 30 m of well bedded sandstones, was deposited in shallow water. These sandstones are commonly ripple marked, and locally contain shell beds with abundant bivalves, gastropods and ammonites. Southwards the formation thins; it is apparently included within 21 m of sandstone in Whitwell on the Hill Oil Borehole and is absent over the Market Weighton Block. South of Market Weighton no rocks of *margaritatus* Zone age are known to the north of Kirton in Lindsey, where there are 5 m of clays with thin ironstones referable to this zone (Figure 6; Plate 5). These beds thicken to 10 m at Lincoln and 20 m at Grantham.

The Cleveland Ironstone is best developed near Guisborough, where it reaches a maximum thickness of 24 m in a contemporaneous east–west syncline. The formation thins southwards, partly due to thinning of the component units, but largely because of unconformable cut-out of lower beds beneath the transgressive Main Seam (*spinatum* Zone). The succeeding sandy shales also contain *Pleuroceras*, but there is evidence that the lowest

Plate 3

1 Hydraulic Limestones near base of Lower Lias, Barnstone Quarry [SK 739 348], Nottinghamshire. Belvoir Castle, on the skyline right of centre, stands on the Middle Lias scarp. (*Sir Peter Kent*)

2 Frodingham Ironstone workings at Winterton Mine, north of Scunthorpe. View looking south [from SE 915 203]. The ironstone is exposed at the bottom of the pit and the Marlstone Rock at the top on the left. The *Pecten* Ironstone is largely hidden beneath slipped debris part way up the workings. The farm, Winterton Grange, lies on the lowest beds of the Lincolnshire Limestone. (L 1688)

Plate 4 *Pecten* Ironstone in Crosby Warren Mine [SE 907 137], north-east of Scunthorpe. The ironstone, about 2.2 m thick, lies within Lower Lias mudstones above the worked Frodingham Ironstone. Note the fallen block, left of centre, with scattered *Pseudopecten* on a bedding plane. (*Sir Peter Kent*)

Plate 5 Marlstone Rock overlying Middle Lias mudstones in quarry [SE 933 005] at Cleatham, near Kirton in Lindsey. The mudstones are worked for cement manufacture. (L 1687)

Plate 6

1 Workings in Ancaster Rag and Freestone at Gregory's Quarry [SK 992 409], Ancaster. The reddened horizon in the limestone marks a stratigraphical break. Overlying Upper Estuarine Beds are exposed in the upper part of the workings. (*Sir Peter Kent*)

2 Kirton Cementstones resting on Santon Oolite in quarry [SE 9450 0157] near Kirton in Lindsey. The Kirton Cementstones consist of a thick hard bed overlying interbedded thin porcellanous concretionary limestones and dark mudstones. The left hand of the observer is resting on the top of the Santon

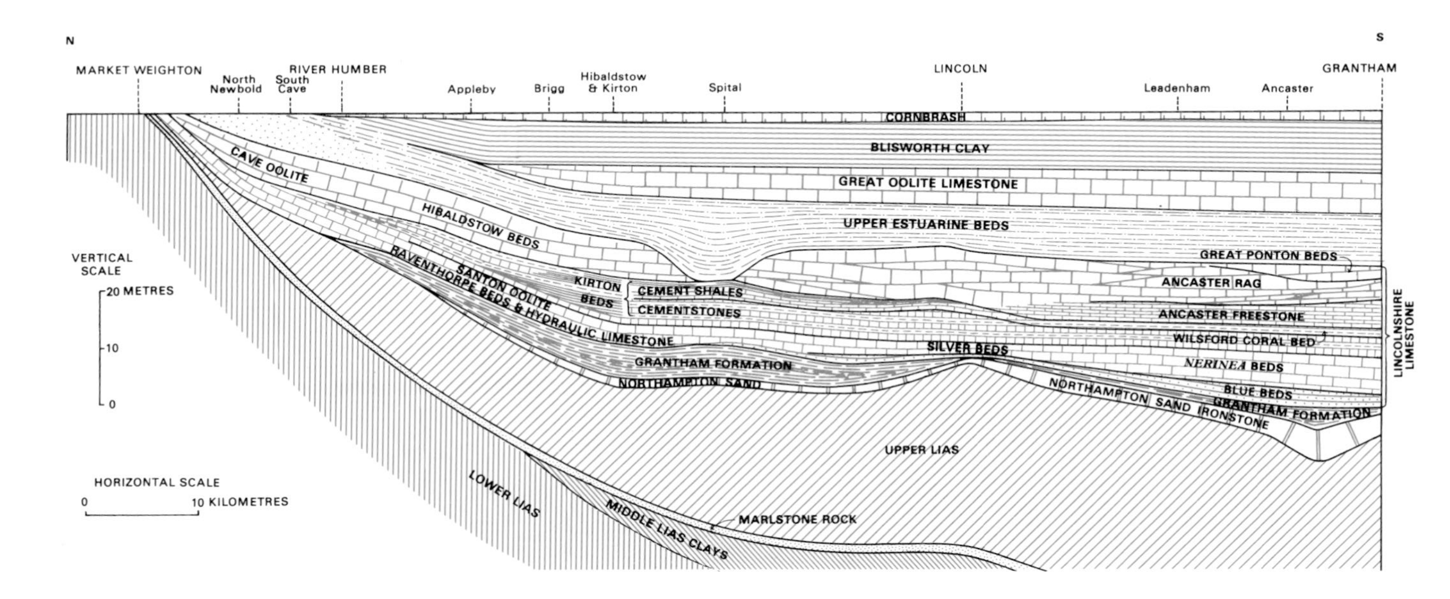

Figure 6 Generalised section of the Middle and Upper Lias and the Middle Jurassic rocks from Grantham to Market Weighton

specimen of *Dactylioceras*, indicating a Toarcian age, was found in the highest bed of the ironstone (Howarth, 1973). Where the thickness of the Cleveland Ironstone is at a maximum the rock is largely a chamosite oolite, and attenuation is associated with passage into siderite mudstone. Deposition was in shallow water and, as in the Frodingham Ironstone, thick-shelled molluscs are common in the oolitic beds.

South of Market Weighton only the *spinatum* Zone element is developed as a calcareous ironstone—the Marlstone Rock (Plate 5). It extends continuously across Lincolnshire except for a 15-km gap at outcrop from Burton past Lincoln to Welbourne. It is too thin and low grade for economic working over most of this area, even where it is excavated as part of the overburden of the Frodingham Ironstone. It is of workable thickness in the Grantham district and has been worked near Caythorpe and at Harston on the region's southern boundary, where it is 2 to 3 m thick and continues into the former Leicestershire ironstone field. The chamosite oolite facies continues eastwards at depth as far as Woodhall Spa and Ruskington on the edge of the fenland, but has not been identified in most offshore wells.

The detailed dating of this shallow-water formation remains to be worked out. The fauna is often extremely rich, but is largely limited to brachiopods (particularly *Lobothyris punctata* and *Tetrarhynchia tetrahedra* (Figure 5) and their allies), and ammonites are very localised. The zonal genus *Pleuroceras* (Plate 7) occurs in the Marlstone Rock in Yorkshire south of Market Weighton and in north Lincolnshire, and *Dactylioceras* occurs in the feather edge of the formation at Welbourne in mid-Lincolnshire and through a metre or so of it at Harston south-west of Grantham. Thus the Marlstone Rock also crosses the Pliensbachian/Toarcian boundary in south and mid-Lincolnshire.

Upper Lias

The Upper Lias of the region, essentially an argillaceous sequence, shows contrasting developments in the northern and southern provinces, for north-east Yorkshire has a much thicker and more varied development than the south. The Toarcian Stage, to which it is assigned, is divided into two sub-stages (Whitbian below and Yeovilian above) of which only the lower is represented over the greater part of the region. Yeovilian rocks, which are widely developed in southern England, were deposited but subsequently eroded over most of the Cleveland Basin; farther south in Yorkshire and in Lincolnshire their absence is probably due partly to erosion and partly to non-deposition. In the region they are limited to a small outcrop east of the Peak Fault on the Yorkshire coast, and to small patches in pre-Dogger synclines in the Cleveland Hills.

In north Yorkshire Upper Lias deposition began with the Grey Shales, some 13.5 m of pale grey silty shales. They contain sideritic nodules and are locally highly pyritic. The finely ribbed *Dactylioceras tenuicostatum* (Plate 8) is common and characteristic, as are large belemnites ('*Belemnites*' *cylindricus*).

The uppermost part of the Grey Shales shows an approach to the lithology of the overlying Jet Rock, which is a richly bituminous shale about 9 m thick, mostly finely laminated and papery weathering, with layers of calcareous concretions (doggers) and, at the top, a thin pyritic limestone (the Top Jet Dogger). In the upper part of the Jet Rock fossil wood converted into tough

Plate 7 Lower and Middle Lias ammonites (natural size)

1 *Schlotheimia angulata* (Schlotheim); *angulata* Zone. **2** *Gagaticeras gagateum* (Young & Bird); *oxynotum* Zone. **3** *Paltechioceras aplanatum* (Hyatt); *raricostatum* Zone. **4** *Aegoceras* (*Androgynoceras*) *maculatum* (Young & Bird); *davoei* Zone. **5** *Amaltheus subnodosus* (Young & Bird); *margaritatus* Zone. **6** *Pleuroceras hawskerense* (Young & Bird); *spinatum* Zone.

black jet is common, and this has provided the basis for the jet jewellery industry of Whitby since Bronze Age times. Ammonites are common in the Jet Rock and may contain liquid petroleum in their chambers, diffused from the surrounding oil-rich shales. Species of *Harpoceras* (Plate 8) are particularly common (including the characteristic *H. falciferum*) with the bivalve *Pseudomytiloides dubius* (Figure 7) in abundance. Well-preserved remains of saurians and of the fish *Lepidotus* have been found in the course of jet mining.

The overlying strata, consisting of the Bituminous Shales, the Alum Shales and the Cement Shales, total some 60 m.

At the top of the 23 m of Bituminous Shales, and included in the *falciferum* Zone, is a double band of pyritic nodules, the *ovatum* Band, in which are occasional masses of siderite mudstone showing cone-in-cone structure and masses of belemnites forming limestones up to 7 cm thick. The nodules contain the ammonite *Ovaticeras ovatum*, peculiar to this band and rare elsewhere in England.

The Alum Shales, belonging to the *bifrons* Zone, consist of up to 37 m of soft grey micaceous shales. The basal 6m, which are almost non-bituminous (the Hard Shales), are capped by a 13-cm band of siderite mudstone. This is succeeded by the so-called Main Alum Shales, 15 m of grey pyritic shales. The upper 13 to 20 m of the Alum Shales, known as the Cement Shales, contain beds of calcareous nodules which were formerly used for making hydraulic cement. The Alum Shales yield large numbers of the ammonites *Dactylioceras commune* and *Hildoceras bifrons* (Plate 8), which find a ready sale to visitors to Whitby; the small bivalve *Nuculana ovum* (Figure 7), belemnites and many other fossils are common. These beds are also famous for the large number of saurians found when the alum quarries were working, representing several species attributed to the genera *Ichthyosaurus* and *Plesiosaurus*, together with the great crocodile *Steneosaurus* and remains of the sturgeon-like fish *Gyrosteus mirabilis*.

The alum industry began in Yorkshire about 1600 AD and continued for some 250 years. Vast excavations were made at Peak, Saltwick, Kettleness, Boulby and elsewhere; the shales were economically quarried near the top of the cliffs where the disposal of the burnt shale presented no difficulties. Inland, alum shale workings occur along the north Cleveland scarp westward to Osmotherley but these openings were of comparatively small importance. The shale was first calcined over a bed of brushwood; the heaps, sometimes reaching a diameter of 30 m with a height of 15 m, burned for a year or more. The burnt shale was then steeped in water in tanks to extract the sulphates of iron and alumina formed by the action of sulphuric acid, derived from the pyrite, on the shale. To the resulting liquor an alkali was added and on evaporation the alum crystallised before the iron salts, which could then be pumped off. How to ascertain the right moment at which to cease heating the liquor, so that the maximum amount of alum had formed while the iron salts were still retained in solution, was an alum maker's 'secret'. It was done by floating an egg in the liquid as a hydrometer.

The upper half of the Upper Lias, the rocks of Yeovilian age, is about 50 m thick on the eastern side of the Peak Fault but is absent on the western side, where the Dogger (Middle Jurassic) rests directly on the Alum Shales. This relationship was formerly ascribed to pre-Dogger uplift and erosion of the

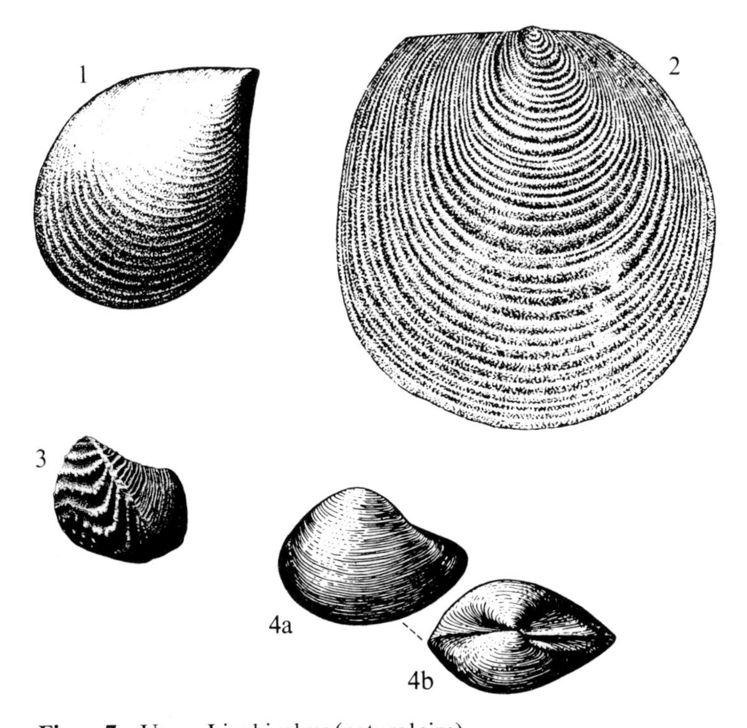

Figure 7 Upper Lias bivalves (natural size)

1 *Pseudomytiloides dubius* (J. de C. Sowerby); **2** *Steinmannia bronni* (Voltz); **3** *Vaugonia pulchella* (L. Agassiz); **4a, b** *Nuculana ovum* (J. de C. Sowerby).

western side, but it is now thought that the fault is a transcurrent one and that contrasting developments from two different parts of the basin have been juxtaposed by lateral movement of several kilometres. Small inland outcrops of the Yeovilian strata are also now known, notably in the Ralph Cross area of the Cleveland Hills.

On the coast, east of the Peak Fault (Figure 8), the Yeovilian beds are overlain with no more than a slight stratigraphical break by the Dogger. Here the Alum Shales are succeeded by about 9 m of dark shales, the Peak Shales, largely obscured by scree and shore boulders, which have yielded species of *Haugia*. They are succeeded by the *striatulus* Shales—16 m of poorly exposed micaceous shales with occasional bands of impure ironstone, from which *Grammoceras striatulum* (Plate 8), *G. thouarsense* and other species have been recorded. The Blea Wyke Beds which follow comprise 13 m of Grey Beds with 8 m of Yellow Beds above. The lower 2 m of the Grey Beds consist of soft grey shales with a basal line of nodules in which *Lingula beani* is abundant; these shales pass up into 7.6 m of grey micaceous sandstones containing *Lingula*, which are followed by 3.6 m of false-bedded sandstones in which nests of the worms *Serpula deplexa* and *S. ? compressa* weather out as burr-like masses. The ammonites *Phlyseogrammoceras dispansum* and *Hudlestonia affinis* also occur in the Grey Beds.

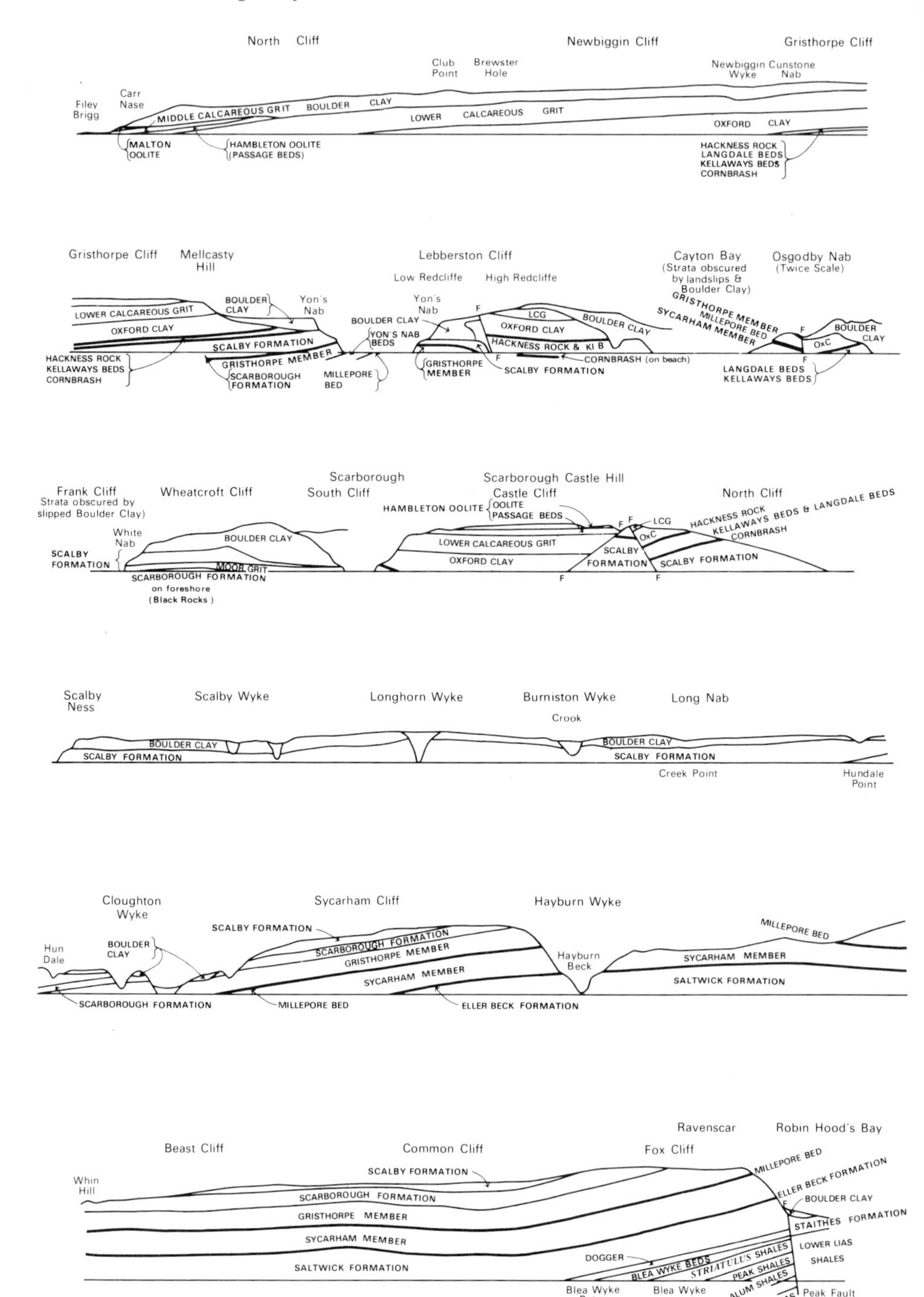

Figure 8 Sections showing the Jurassic rocks and Drift deposits along the cliffs from Filey to Robin Hood's Bay

Plate 8 Upper Lias ammonites (natural size)

1 *Dactylioceras* (*Orthodactylites*) *tenuicostatum* (Young & Bird); *tenuicostatum* Zone. **2** *Harpoceras exaratum* (Young & Bird); *falciferum* Zone. **3** *Hildoceras bifrons* (Bruguière); *bifrons* Zone. **4** *Dactylioceras commune* (J. Sowerby); *bifrons* Zone. **5** *Grammoceras striatulum* (J. de C. Sowerby); *thouarsense* Zone.

The Yellow Beds are soft massive yellow sandstones with occasional calcareous concretions and abundant worm tubes. They are 8 m thick at Blea Wyke but thin to 2.7 m at the Peak Fault. The finely ribbed ammonite *Dumortieria moorei* has been recorded from the upper part, and on this evidence the beds are assigned to the *moorei* Subzone of the *levesquei* Zone. The top of these beds is hardened and bored, and on this surface at Blea Wyke Point lies an irregular 45-cm *Terebratula* Bed, which at the Peak Fault is represented by a fossiliferous 30-cm pebble bed. For many years it has been known that the Lower/Middle Jurassic junction occurs at a non-sequence at Blea Wyke and the work of Rastall and Hemingway has shown that this non-sequence is immediately below the *Terebratula* Bed.

In contrast, the Upper Lias of southernmost Yorkshire and of Lincolnshire is much more uniform, consisting almost entirely of mudstones and shales. The lower part of the Upper Lias consists of grey mudstones with pyrite, overlain by paper shales. In the Grantham district the basal beds probably belong to the *falciferum* Zone, but farther north they include the upper subzones of the *tenuicostatum* Zone. Near Lincoln the basal metre consists of greenish shale with, near the top, *Tiltoniceras antiquum* and numerous mainly fine-ribbed dactylioceratids. This has been called the Transition Bed and is referred to the *semicelatum* Subzone. In north Lincolnshire and south Yorkshire recent boreholes have shown that the basal strata extend down to the *clevelandicum* Subzone. The base of the Upper Lias is thus clearly seen to be diachronous.

The upper part of the preserved Upper Lias in Lincolnshire and south Yorkshire consists of soft shales with sporadic lines of limestone nodules, belonging to the higher part of the *falciferum* Zone and to the *bifrons* Zone. In the unweathered state the shales yield numerous ammonites, predominantly of the genera *Harpoceras* and *Dactylioceras*. Notable horizons are a nodular limestone a little above the highest paper shales with *Harpoceras* and an abundance of the small pelagic gastropod *Coelodiscus*, known from the Grantham district, near Lincoln (Burton) and at Kirton in Lindsey, and, high in the shales at Welton and Spital, a shelly bed with common *Vaugonia pulchella* (Figure 7) and other molluscs.

In regional terms the Toarcian zones thin progressively southwards, rocks of the higher zones being present beneath the transgressive Northampton Sand in the south. The detailed trends are not yet understood, but are certainly less regular than this, for the *fibulatum* Subzone of the *bifrons* Zone, which is overlapped by Middle Jurassic rocks north of Grantham, is locally seen again north of Lincoln, and in the former area the overall thickness of the Upper Lias varies between 30 and 60 m along the strike within short distances.

Lias—regional thickness changes

There is more reliable regional information on total Lias thicknesses (Figure 9) than that of other Jurassic divisions, particularly beneath the North Sea.

The East Midlands Shelf is overlain by 250 to 300 m of Lias, with minor thickness variations in the central area. There is steady attenuation northwards towards the Market Weighton Block and south-eastwards towards the East Anglian Massif. Sharp thickness changes occur in Yorkshire north of Market Weighton, for which contemporaneous faulting may be responsible. There are nearly 500 m in the Locton Gasfield in the Cleveland Basin and thicknesses are not much less on the northern edge of the Cleveland Hills (Tocketts), but a regional northward attenuation beyond this is shown by offshore evidence. Otherwise north Yorkshire shows dominantly westerly thinning of the Lias. Beyond the eastern edge of the East Midlands Shelf, some 40 km E of the Holderness and Lincolnshire coasts, there is a sharp thickness increase into the north-west–south-east Sole Pit Trough, where 900 m of Lias were penetrated in the West Sole Gasfield. This thick offshore sequence accumulated in a contemporaneous syncline which was later subject to inversion (pp.114–115). It is essentially argillaceous, although thick sand beds are reported in the Upper Lias near the eastern edge of the shelf.

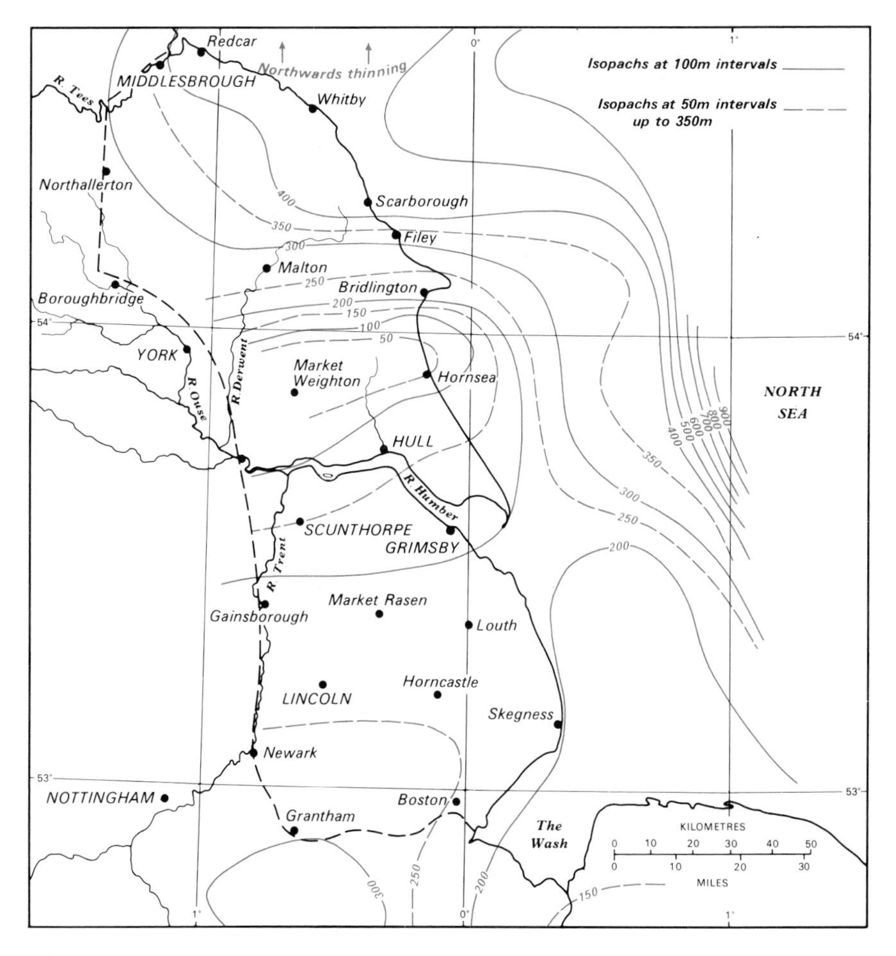

Figure 9 Thickness variations in the Lower Jurassic rocks

6. Middle Jurassic

The Middle Jurassic provides some of the most important feature-forming rocks in the region. Their topographic expression in turn reflects the variation in the regional structure. In the south the eastward-dipping Lincolnshire Limestone forms a simple scarp, Lincoln Edge (Figure 2), extending from Grantham to the Humber and broken only by river gaps: in contrast, the faulted and gently folded limestones of the Tabular and Howardian Hills (Middle Jurassic and Corallian) provide a relatively complex outcrop, and the deltaic sandstones farther north form the broad moorlands which extend from the Cleveland Hills to the coast.

The stratigraphy is complicated, for the region includes the greater part of the transition from the fully marine development of southern England to a dominantly fresh-water regime in the Cleveland Basin. The change begins in the south Midlands with the incoming of the 'Estuarine Beds' in north Oxfordshire and Northamptonshire; the Lincolnshire Limestone changes progressively northwards across Lincolnshire and south Yorkshire from limestones deposited in clear water to a more muddy and silty facies; the Great Oolite Limestone thins northwards and disappears before reaching the Humber. In terms of overall thickness (Figure 10), however, the northward limitation of the marine strata is more than offset by the incoming of non-marine beds, particularly north of the Market Weighton area, for during Middle Jurassic times north-east Yorkshire was occupied by an extensive delta. Here thick sands were laid down, and limestone is restricted to intercalations representing occasional periods of marine flooding.

The nomenclature, in north Yorkshire especially, has always been complex because of discontinuous strata and lateral changes in facies, and it has become further complicated by adjustments to meet changing concepts of classification and sedimentology. Although the deltaic nature of the northern beds was recognised by Simpson in 1868, the name 'Estuarine' with its subdivisions was universally adopted in 1880 and still appears on many geological maps, so that for reference it is here retained at the side of newer terms. Various authors stressed the truly deltaic nature of the beds (eg Black, 1929; Wilson, 1948), and Hemingway (1949) revised the subdivisional names in deltaic terms. This usage has since been overtaken by modern lithostratigraphic names (Hemingway and Knox, 1973) (Table 4).

South Lincolnshire

The Middle Jurassic sequence (Figure 6) begins with the Northampton Sand Ironstone, which lies with a slight angular unconformity on the Upper Lias. South of Lincoln it consists of siderite mudstones, sandstones and limestones, chamosite oolite and limonite oolite, and is alternatively known as the Northampton Ironstone. It is a low-grade siliceous iron ore, green when fresh, which weathers to rich brown limonite with a characteristic 'box-stone'

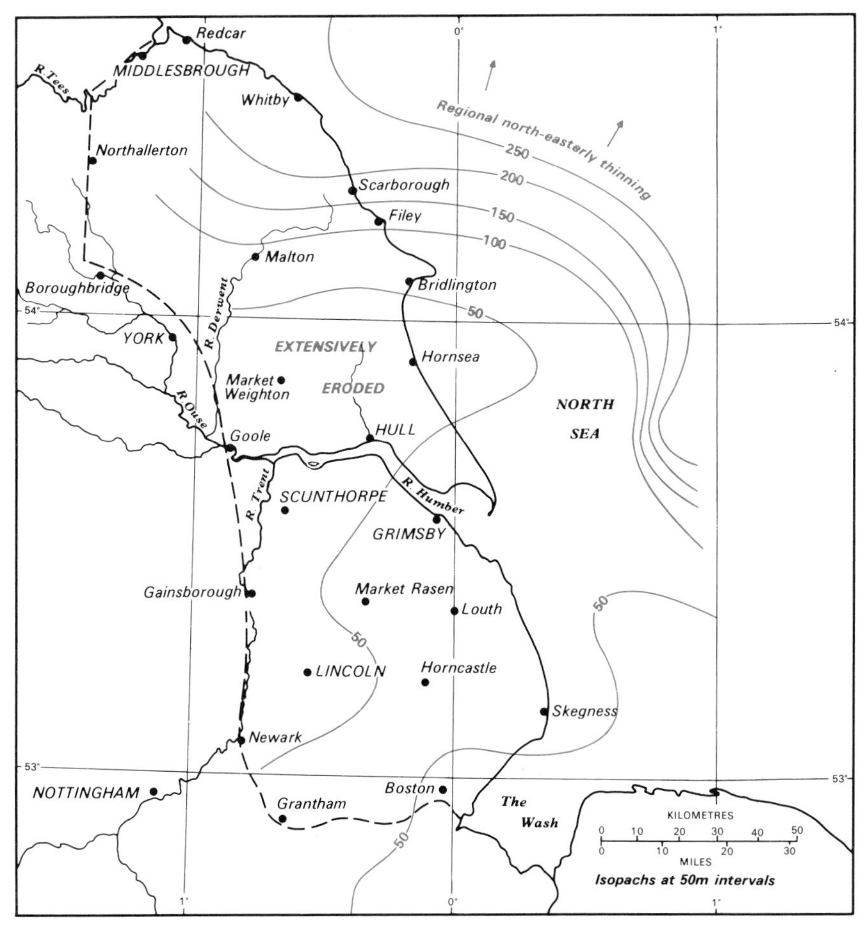

Figure 10 Thickness variations in the Middle Jurassic rocks

structure. The overall thickness is 7 to 8 m, of which only a part (3 to 7 m) has been worked. The most extensive workings lie south of the region's boundary in Leicestershire and Northamptonshire, but they have now been phased out north of the River Welland in favour of richer imported ores.

Fossils in the formation are localised but sometimes abundant; they include numerous bivalves, brachiopods, notably *Lobothyris trilineata* (Plate 9), and rare ammonites including several specimens of *Leioceras opalinum* found at Harlaxton near Grantham, which establish that the Northampton Sand belongs to the basal zone of the Bajocian.

The formation thins rapidly eastwards down-dip, and beneath the Fens is absent or represented only by a line of ironstone nodules. The marine connection with the equivalent beds in southern England thus probably lay to the south-west.

In Lincolnshire the Northampton Sand is conformably overlain by the Grantham Formation ('Lower Estuarine Series'), which consists of fine sands and clays with low-grade ironstone, 2 to 8 m thick. Part of this sequence consists of marsh clays, and vertical plant remains are common in both sands and clays. Ironstone occurs, particularly in the basal sands, and has occasionally been worth taking with the underlying Northampton Sand Ironstone. In an area extending eastwards from the Leicestershire border through Grantham and Sleaford and across the Fens a 1-m marine shale is developed in the middle part; this has yielded a bivalve fauna with common *Vaugonia v-costata* and other species, *Gervillella sp.*, *Modiolus sp.*, small pectinids, '*Nuculana*' *sp.*, the brachiopod *Lingula kestevenensis* and a single specimen of a button coral *Chomatoseris* [*Anabacia*]. Floods of small bivalves similar to '*Corbula*' occur in the shales above. The marine element disappears southwards across Rutland, but the fauna becomes richer westwards and there is a presumption that (as with the Northampton Sand) open sea lay to the west of the present outcrops (Kent, 1975).

This first 'estuarine' episode was followed by a marine transgression which resulted in the deposition of the main element of the Middle Jurassic succession, the Lincolnshire Limestone. This formation thins to the south and east from a maximum of 43 m in the Grantham district to feather edges off the Lincolnshire coast and in the southern Fens. Northwards into Yorkshire it is cut out by the sub-Cretaceous unconformity at Market Weighton.

In south Lincolnshire it consists almost entirely of limestone, developed in a shallow-water facies. The Blue Beds form the basal part of the Lincolnshire Limestone over most of its outcrop; they are very sandy ferruginous blue-hearted limestones 1 to 2 m thick. The fauna is limited, with the bivalves *Gervillella acuta*, *Propeamussium* cf. *laeviradiatum* and *Pinna cuneata* common.

The Blue Beds are succeeded by well-bedded oolitic limestones which have been called the *Nerinea* Beds from the characteristic gastropod (*N. cotteswoldiae*) which occurs in large numbers, along with species of *Trigonia*, *Pinna* and compound corals. Above them is a sequence of thin-bedded micritic limestones with clay partings, the southerly development of the Kirton Cementstones of north Lincolnshire. These beds contain a limited macrofauna, with common '*Lucina*' *wrighti*, *Pinna cuneata*, *Lima rodburghensis* and *Pholadomya sp.*, but with a particularly rich assemblage of ostrocods (Bate, 1967). In the upper part of these beds the small spinose rhynchonellid, *Acanthothiris crossi* (Plate 9) is common, especially in the topmost limestone—the *crossi* Bed. This bed forms one of the best markers in the Lincolnshire Limestone and occurs with only minor interruptions northwards across Lincolnshire almost to the Humber, separating the relatively variable Lower Lincolnshire Limestone from the more homogeneous Upper Lincolnshire Limestone.

The *crossi* Bed is underlain by shaly beds with large corals, the Wilsford Coral Bed, which extends as far north as Heighington near Lincoln. *Zeilleria wilsfordensis* is a characteristic brachiopod.

Above the *crossi* Bed are coarsely oolitic limestones with rare ammonites. These are followed by and interdigitate with coarser 'rags'—calcarenites and calcirudites, consisting largely of small and fragmentary fossils. In this sequence the thicker bedded and more uniform limestones, often with a sparry matrix, provide high-grade building stones—the freestones of Ancaster (Plate 6.1) and, to the south of the region, those of Clipsham, Ketton and Weldon.

Where the Lincolnshire Limestone is at its thickest, in the Grantham district, the highest beds include some 5 m of limestones crowded with gastropods and finally 7 m of limestones with common terebratulids, which have been named the Great Ponton Beds and which may be in part younger than the 'rags' of Ancaster and Clipsham. Brachiopods of the genus *Weldonithyris* are present in both subdivisions at Ponton.

Ammonites have been found in this area from the base of the *Nerinea* Beds to above the *crossi* Bed. Below the latter they are all referable to the *discites* Zone and above it to the *laeviuscula* Zone. However, the zonal position of the highest Lincolnshire Limestone remains, in the absence of ammonites, uncertain.

The widespread unconformity present within the Lincolnshire Limestone in the Kettering area, where the Upper Lincolnshire Limestone rests on a channelled surface of the older beds, is less marked in this region, although it is well seen at Ancaster, where the *crossi* Bed and some of the underlying strata are cut out. When deposition of the Lincolnshire Limestone ceased the area was subject to gentle folding and widespread erosion. No limestones equivalent to the Scarborough Limestone (*humphriesianum* Zone) of north Yorkshire are known in Lincolnshire, but they may have been deposited and since removed.

When deposition was resumed after the erosional interval it was clastic and non-marine: the Upper Estuarine Beds began with a fresh-water phase, with deposition of coloured clays and silts which include a coaly layer useful as a marker. Pale sands are occasionally present above the persistent basal 'Ironstone Junction Band'. The fresh-water beds, 2 to 3 m thick, were followed by rhythmic units (five have been recognised) in each of which a marine layer, sometimes a limestone, is followed by brackish and fresh-water, greenish and purple clays, making up a total thickness of 8 to 10 m.

Viviparus has been recorded in these fresh-water beds near Lincoln, and the brachiopod *Lingula kestevenensis* characterises one of the lower marine levels. The termination of the sequence was marine, with dark shales containing *Liostrea hebridica* and *Kallirhynchia sharpi* (Plate 10).

The Upper Estuarine Beds are overlain by the Great Oolite Limestone, which corresponds in facies and position (although not precisely in age) with the White Limestone of Oxfordshire and the south. It is a 7 to 8-m sequence of pale soft limestone with marly partings which forms only a minor feature on the landscape. The fauna is largely limited to bivalves. The oyster *Liostrea hebridica* occurs in large numbers and is accompanied by *Modiolus imbricatus*, *Pholadomya* and other bivalves: brachiopods are locally common and in this area are dominated by rhynchonellids, with the characteristic *Kallirhynchia sharpi* in the basal beds.

Overlying the Great Oolite Limestone is the Blisworth Clay, deposited under marine or brackish conditions and consisting of dark green and grey clays with abundant oysters. *Liostrea hebridica* s.s. continues in rock-forming quantity and is accompanied by the subspecies *L. h. subrugulosa*. The formation is persistent and uniform, with a thickness of 7 to 10 m.

The final formation of this varied group of strata is the Cornbrash, which consists of detrital limestones, with marly and sandy partings, measuring only 2 to 3 m in thickness. This lies across the Middle/Upper Jurassic boundary. It is usually richly fossiliferous with abundant brachiopods and bivalves, and less frequent but not uncommon ammonites.

In this area the Lower Cornbrash is somewhat impersistent, but the characteristic ammonite *Clydoniceras discus* and the brachiopods *Cererithyris intermedia* (Plate 10) and *Obovothyris obovata* have been found at a number of places. The Upper Cornbrash, which is of Upper Jurassic (Callovian) age, contains macrocephalitid ammonites, the typical brachiopods *Microthyridina* and *Rhynchonelloidella* and the oysters *Liostrea undosa* and *Lopha marshii* (Plate 10). It continues across the whole area, and is frequently seen in ditch sections near the edge of the Fens.

North Lincolnshire

Between Lincoln and the Humber the succession (Figure 6) is in many respects intermediate between the successions of the Midlands and the Cleveland Basin, with significant lithological changes in all the formations.

The Northampton Sand continues northwards to Scunthorpe and dies out near Winterton. It is up to 3.5 m thick, but locally thins to 1 m or less. At Wressell it measures 1 m and contains a marine fauna which links it with the Dogger of north Yorkshire, including *Homoeorhynchia* cf. *cynica*, *Tetrarhynchia spp.*, *Lobothyris trilineata* (Plate 9) and *Eopecten gradus*. Throughout most of the area the formation is a calcareous ferruginous sandstone, lacking the workable iron content of the area to the south.

The Grantham Formation ('Lower Estuarine Series') is absent in the Lincoln area but elsewhere is developed as white or lilac-coloured sands with beds of carbonaceous clay, to a maximum thickness of 7 to 8 m along the outcrop from Fillingham northwards. The formation is still represented by thick sands near Scunthorpe, but thins out towards the Humber, and the beds formerly mapped as 'Lower Estuarine Series' in south Yorkshire have been shown by boreholes to be a basal argillaceous facies of the Lower Lincolnshire Limestone (see below).

The Lincolnshire Limestone shows a progressive northward increase in terrigenous clastics, best seen in the Kirton Cementstones and Kirton Shale. The Kirton Shale, up to 3 m thick, contains large masses of corals and is the equivalent of the Wilsford Coral Bed of south Lincolnshire.

The basal beds continue as blue-hearted sandy limestones at least as far as Cameringham and Glentworth, but near the latter a dense, pale limestone with thin shells appears immediately above them; this continues northwards as the Hydraulic Limestone. Sandy limestones, sands and silts, named the Raventhorpe Beds, are developed above the Hydraulic Limestone.

Bedded oolites of the Lower Lincolnshire Limestone, the 'Silver Beds', equivalent to the *Nerinea* Beds of the south, continue north of Lincoln into the Caenby and Atterby area, and farther north are known as the Santon Oolite. The overlying Kirton Cementstones (Plate 6.2) are dark fine-grained limestones, partly sandy and commonly with scattered irregular black ooliths, interbedded with shale. The fauna is dominated by bivalves, among which *Trigonia hemisphaerica* is common.

The Upper Lincolnshire Limestone continues northwards as a shallow-water oolite and calcarenite, the Hibaldstow Beds, some 10 m thick where fully developed, which resemble the freestones and rags of south Lincolnshire and Rutland.

Ammonites found in the 'Blue and Silver Beds' near Lincoln (Greetwell) are species of *Hyperlioceras* indicative of the *discites* Zone.

The contact of the Lincolnshire Limestone with the overlying Upper Estuarine Beds in this area is a low-angled unconformity. Thus the Hibaldstow Beds, which are thick near Nettleham and Hibaldstow, are cut out completely near Spital in the Street and again at Snitterby, so that the overall thickness of the limestone in the area varies from about 15 to 23 m.

The overlying Upper Estuarine Beds are not well known, but appear to be less marine than in south Lincolnshire. Generally the succession consists of pale sands in the lower part with overlying mauve and greenish clays, pale silts and sands with impersistent coaly beds. Limestone was observed in sections near Spital, and the brachiopod *Kallirhynchia* was seen at the contact with the overlying Great Oolite Limestone. Recent boreholes astride the Humber show traces of marine and brackish elements.

The Great Oolite Limestone continues northwards beyond Lincoln, thinning to a feather edge near Brigg. The fauna is variable; brachiopods have been found near Lincoln (Sudbrook), but more commonly the fauna is limited to shallow-water bivalves including *Liostrea hebridica* with occasional rhynchonellids.

The Blisworth Clay similarly persists into north Lincolnshire with little apparent change; the oysters *L. hebridica* s.s. and *L. h. subrugulosa* remain characteristic, and the thickness is maintained at 5 to 7 m. It extends a little beyond the limit of the Great Oolite Limestone, but is not recognisable in boreholes near the Humber.

The Cornbrash continues through north Lincolnshire as a fully marine unit, consisting mainly of detrital limestone, but locally containing shelly clays. As in south Lincolnshire the Upper Cornbrash is the more persistent part, with the Lower Cornbrash recognised locally.

Humber to Market Weighton

In south Yorkshire the Middle Jurassic rocks (Figure 6) consist of a lower marine sequence equivalent to the Lincolnshire Limestone and an upper, mainly non-marine, sequence, the Upper Estuarine Beds. No equivalents of the Northampton Sand and Grantham Formation of north Lincolnshire have been recognised, but the Cornbrash reaches just across the Humber, the feather edge of the Upper Cornbrash being present near North Ferriby.

The strata originally mapped as Lower Estuarine Beds in south Yorkshire are largely marine, consisting of ferruginous silts, limestones and shales with marine fossils. They are the equivalents of the lower part of the Lower Lincolnshire Limestone.

They have been penetrated in the IGS Humber boreholes and are also well developed at Brantingham; relics beneath the Chalk farther north (eg at Sancton and Millington) indicate that they formerly extended across the Market Weighton area.

The main part of the marine sequence of the Middle Jurassic in south Yorkshire is the Cave Oolite, an oolitic limestone, the upper part of which is in stratigraphical continuity with the Hibaldstow Beds of the Upper Lincolnshire Limestone. The lowest beds contain the 'millepore' *Haploecia straminea*

(Plate 9) and have yielded small *Acanthothiris* at South Cave, Givendale and in the Humber boreholes. The oolite beds above are coral-bearing, and *Hyperlioceras* has been found in their lower part at South Cave (Senior and Earland-Bennett, 1973).

In south Yorkshire the Upper Estuarine Beds are well developed as sands, up to 14 m thick with overlying northward-thinning mudstones. It is uncertain whether the formation continues as far as Sancton (as formerly mapped); it may have been overlapped by the Kellaways Beds between there and North Cave.

Cleveland Basin

Although the transition from the largely marine development of Lincolnshire into the northern deltaic facies is to some extent gradational, the sub-Cretaceous erosional interruption across the Market Weighton Block and the sudden expansion of the various Middle Jurassic units (particularly the non-marine parts) in the basin to the north give the Cleveland area a very distinctive character. The only comprehensive account of the stratigraphy is still that of Fox-Strangways (1892), but details have been added by later authors, and the petrography, depositional environment and palaeobotany are well documented.

Table 4 Classification of the Middle Jurassic rocks of north-east Yorkshire

Fox-Strangways, 1892*			Hemingway, 1949		Hemingway and Knox, 1973	
Upper Estuarine Series			Upper Deltaic Series		Scalby Formation	
Scarborough or Grey Limestone Series			Grey Limestone Series		Scarborough Formation	
Middle Estuarine Series			Middle Deltaic Series	Gristhorpe Sub-Series	Cloughton Formation	Gristhorpe Member
Millepore Series				Millepore Sub-Series		Lebberston Member
Lower Estuarine Series				Sycarham Sub-Series		Sycarham Member
	Eller Beck Bed	Hydraulic Limestone	Eller Beck Bed	Hydraulic Limestone		Blowgill Member†
					Eller Beck Formation	
			Lower Deltaic Series		Saltwick Formation	
Dogger			Dogger		Dogger Formation	

* Following Geological Survey, 1880
† Including Hydraulic Limestone

The basal unit of the Middle Jurassic of the Cleveland Basin is the Dogger, a highly variable marine formation comprising conglomerates, sandstones, shales, limestones and ironstones. 'Dogger' is a name commonly used for hard rounded stones, often concretionary, and appears to have been given to this formation in the alum works from its weathering habit. It is a name which was applied (mistakenly) to the Middle Jurassic generally across Europe, whereas in the English and original sense it is restricted as a proper name to the ferruginous strata at the base.

Over much of the area the Dogger is a calcareous or chamositic sandstone, 1 to 2 m thick, but on the coast at Blea Wyke near Robin Hood's Bay (Figure 8) it expands to 12 m with rich shelly beds. The latter include the highly fossiliferous *Nerinea* Bed with *N. cingenda* (Plate 9), bivalves and corals near the top, and the *Terebratula* Bed with *Lobothyris trilineata* (Plate 9) near the base, containing gastropods and corals. Across the Peak Fault from Blea Wyke the thickness is reduced; the *Terebratula* Bed is replaced by conglomerate and the chamositic development by sideritic sandstones. Near Whitby the Dogger has yielded ammonites of the *opalinum* Zone (basal Bajocian) (Figure 11) and is therefore comparable in age with the Northampton Sand. At Loftus, however, and in the hills west of Bilsdale ammonites indicate the later *murchisonae* Zone. Pebbles (often fossiliferous) are present in the lower part at many localities: this is a widespread feature, symptomatic of the extensive erosion of the Lias which preceded Dogger deposition; it is interesting that on the evidence of the derived fossils this erosion reached the Lower Lias near Saltwick Bay.

Inland at Rosedale there was a notable development of iron ore, now worked out, at this horizon. This was dark blue magnetic oolitic ironstone

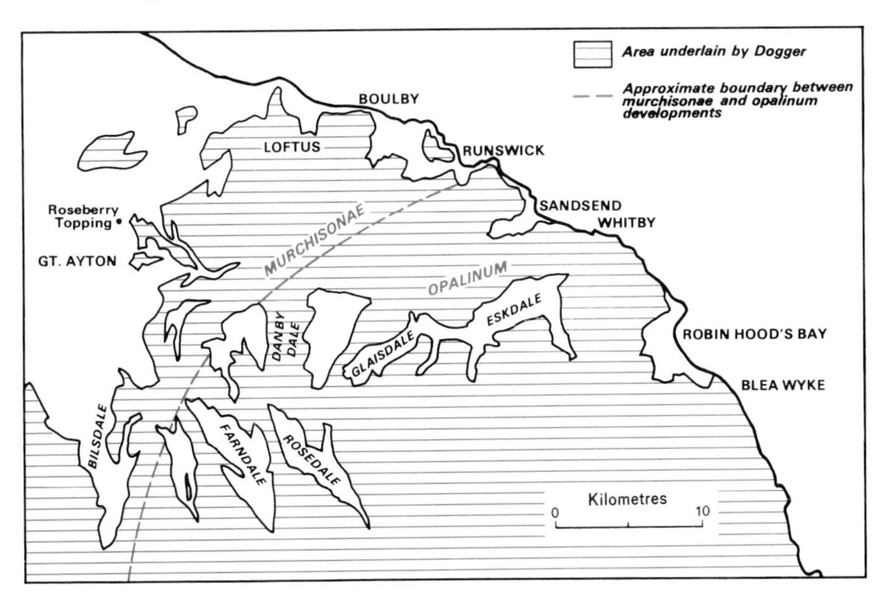

Figure 11 Geographical variation in the zonal development of the Dogger in north-eastern Yorkshire

deposited in a channel system to a maximum thickness of 20 m, though much of the worked ironstone was thinner than this. The rock was locally very fossiliferous, some parts being rich in echinoids (*Acrosalenia*). The same facies occurs offshore on the south-east flank of the Scarborough Dome, but its precise relationships to the Upper Lias are not known (Dingle, 1971).

In the western part of the Cleveland Hills and in the Howardian Hills the Dogger continues as a ferruginous sandstone or limestone, sometimes shelly, ranging in thickness from 5 m to less than 1 m over short distances. It is well exposed in the Derwent Valley at Kirkham, and near Castle Howard a bed, apparently the Dogger, is developed a short distance below the Hydraulic Limestone (as is the Northampton Sand in north Lincolnshire). The Dogger is exposed in Leavening Beck and has been proved in the IGS Acklam boreholes, near where it thins out southwards on the flank of the Market Weighton Block.

Above the Dogger comes the first of the major non-marine interludes, represented by the Saltwick Formation, formerly the lower part of the Lower Estuarine Series or the whole of the Lower Deltaic Series, deposited when rivers from the north spread great thicknesses of sand and mud in a network of channels and marsh flats. The strata are essentially delta top-set beds, the facies being much the same as that of the Coal Measures, although coals are less frequent and of poor quality; two seams have, however, been worked. The shales in this dominantly sandstone sequence (see Plate 14.1) are sometimes rich in plants, as seen, for example, in the '*Thinnfeldia*' Leaf Bed of the Roseberry Topping outlier near Great Ayton. Leaves of *Pagiophyllum*, *Maratiopsis* and *Zamites* are particularly characteristic in different places, the species indicating Liassic affinities, and the total flora is a very rich one, extensively studied. The horsetail *Equisetum* is widespread, often in growth position.

The fresh-water character of the formation is emphasised by the occurrence of *Unio* beds (see Plate 9), as at Brow Alum Quarry.

Marine waters subsequently flooded into the delta to deposit the Eller Beck Formation (named after Eller Beck in Goathland). The formation is widespread and, although only 4.5 to 8 m thick, forms a good marker. The lower part consists of shales with a basal sideritic ironstone, locally oolitic; the upper part is a shaly sandstone. There is a large fauna mostly of bivalves; *Astarte minima*, *Gervillella*, *Liostrea*, *Pholadomya* and various limids are characteristic. Open sea is believed to have lain to the east.

The next marine transgression, represented by the Blowgill Member of the Cloughton Formation, a few metres above the Eller Beck Formation, has a

Plate 9 Middle Jurassic (Bajocian) fossils (opposite)

1 *Sonninia* (*Euhoploceras*) *acanthodes* S. S. Buckman, $\times \frac{1}{2}$; Lincolnshire Limestone. **2** *Pteroperna plana* Morris & Lycett, $\times \frac{1}{2}$; Scarborough Formation. **3** *Megateuthis aalensis* Voltz, $\times \frac{1}{2}$; Scarborough Formation. **4** *Parvirhynchia kirtonensis* Muir-Wood, $\times 4$; Lincolnshire Limestone. **5** *Unio kendalli* Jackson, $\times \frac{1}{2}$; Saltwick Formation. **6a, b** *Dorsetensia sp.*, $\times 1$; with adherent *Meleagrinella*; Scarborough Formation. **7** *Lobothyris trilineata* (Young & Bird), $\times 1$; Dogger. **8a, b** *Acanthothiris crossi* (Walker), $\times 4$; Lincolnshire Limestone. **9** *Plagiostoma rodburghensis* (Whidborne), $\times \frac{1}{2}$; Lincolnshire Limestone. **10** *Leioceras sp.*, $\times 1$; Northampton Sand Ironstone. **11** *Nerinea cingenda* (Phillips), $\times 1$; Dogger. **12** *Haploecia straminea* (Phillips), $\times 2$; probably Millepore Bed.

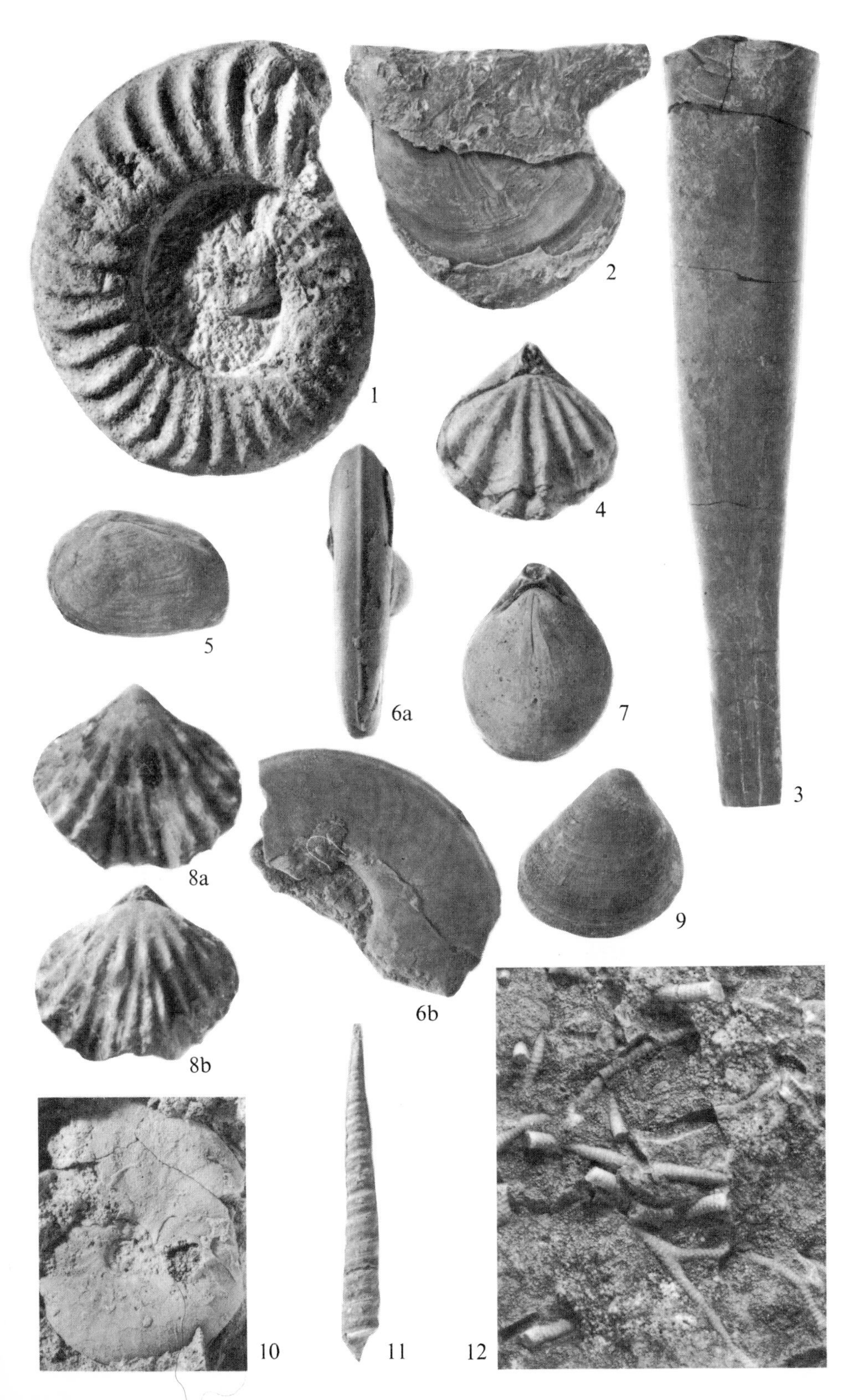
1
2
3
4
5
6a
7
8a
6b
9
8b
10
11
12

different distribution to the latter, being best developed in the south-west of the area. It appears north of Garrowby and reaches a maximum of about 7 m in the Howardian Hills, where it includes marine shales, sandstones and limestones, including the Hydraulic Limestone. Two levels are sideritic but do not contain enough iron to be termed ironstones. The Hydraulic Limestone is a smooth-fracturing argillaceous limestone, generally less than 1 m, but up to 2.2 m, thick. The fauna is poor, mostly of thin-shelled bivalves; it includes *Liostrea*, *Pinna*, *Meleagrinella*, *Camptonectes* and limids, together with ostracods which have led to a correlation with part of the Kirton Cementstones of north Lincolnshire.

The overlying Sycarham Member (formerly the upper part of the Lower Estuarine Series or the lower part of the Middle Deltaic Series), up to 50 m thick, is shalier than the Saltwick Formation but similarly contains thick current-bedded sandstones. Filled channels are infrequent. Plant remains occur—mainly fragments of conifer leaves and spores—but they are fragmentary and clearly drifted, more than at most other levels in the deltaic beds.

When the sea flooded over the delta again it deposited the Lebberston Member. It was a northward extension of the shallow, lime-rich sea in which the Cave Oolite and Hibaldstow Beds were deposited. In the east the lower part of the Lebberston Member is traditionally named the Millepore Bed (Figure 8), the equivalent of which in the west, in the Howardian Hills, is the Whitwell Oolite. The Millepore Bed takes its name from the wide abundance of the bryozoan ('millepore') *Haploecia straminea* (Plate 9). The basal part is sandy, particularly in the west, but it is largely oolitic. In the Cleveland Hills the calcareous oolites give place northwards to a sandy ferruginous facies—sideritic sandstones and mudstones—and beyond the latitude of Whitby change further into non-marine sands, silts and clays.

The upper, regressive, part of the Lebberston Member is called the Yons Nab Beds (Figure 8): these are shales and sandstones, mudstones and sandy limestones up to 9 m thick, with *Trigonia* and other bivalves. The south-westerly equivalent of these beds is the Upper Limestone, which consists of calcareous sandstones, limestones and marls overlying the Whitwell Oolite.

Deltaic conditions were again resumed with deposition of the Gristhorpe Member (formerly Middle Estuarine Series or the upper part of the Middle Deltaic Series). Again there was a regime of marsh flats and shifting river channels, its deposits inseparable from those of the Sycarham Member where the intervening marine strata are absent. The thickness is up to 30 m. Coals in these beds were widely worked in bygone years, and most of the two thousand old coal pits on the moors are said to have been excavated in the Gristhorpe Member. The Gristhorpe Plant Bed has yielded a particularly rich flora

Plate 10 Middle Jurassic (Bathonian) and Upper Cornbrash fossils (natural size, except 3, 8 and 9) (opposite)

1 *Trigonia elongata* J. de C. Sowerby; Upper Cornbrash. **2** *Cererithyris intermedia* (J. Sowerby); Lower Cornbrash. **3** *Lopha marshii* (J. Sowerby), $\times \frac{1}{2}$; Upper Cornbrash. **4a, b** *Microthyridina siddingtonensis* (Davidson); Upper Cornbrash. **5** *Modiolus bipartitus* J. Sowerby; Upper Cornbrash. **6** *Liostrea hebridica* (Forbes); Blisworth Clay. **7** *Trigonia rolandi* Cross; Lower Cornbrash. **8** *Kallirhynchia sharpi* Muir-Wood, $\times 1\frac{1}{2}$; Great Oolite Limestone. **9a, b** *Macrocephalites* (*Dolikephalites*) *dolius* (S. S. Buckman), $\times \frac{1}{2}$; Upper Cornbrash.

1
2
3
4a
4b
5
6
7
8
9a
9b

including Bennettitales, Ginkgoales, conifers and ferns; it is particularly interesting for the occurrence of *Caytoniales*—formerly thought to be the earliest angiosperm.

The fourth Middle Jurassic marine incursion into the Cleveland Basin led to deposition of the Scarborough Formation (formerly the Grey Limestone Series). This formation is absent in south Yorkshire and Lincolnshire, and although it has not yet been reported from the North Sea it is now thought, on the evidence of faunal linkage with Germany, that deposition was due to a transgression from the east. In general terms the Scarborough Formation consists of limestone, sandstone and shale in upward sequence, but there is some evidence of cyclic deposition. On the western escarpment near Brandsby it consists of flaggy limestones and calcareous sandstones (the Brandsby Roadstone) and a crinoidal limonitic sandstone with large rounded quartz pebbles in the north (the Crinoid Grit).

Particular interest attaches to the fauna of the Scarborough Formation, which includes abundant *Meleagrinella* [*Pseudomonotis*] *lycetti*, less frequent *Trigonia signata* and ammonites such as *Dorsetensia*, various stephanoceratids including *Stephanoceras* aff. *gibbosa*, *S.* aff. *crassicostatum*, *S.* cf. *zieteni*, *S.* aff. *triptolemus* and *Teloceras blagdeni*. It appears that all three subzones of the *humphriesianum* Zone are present (Parsons, 1977). The total faunal list is a long one, including echinoids, brachiopods, ostracods, numerous bivalve and gastropod species as well as ammonites and belemnites (Plate 9).

The final deltaic episode was that of the Scalby Formation (formerly Upper Estuarine Series or Upper Deltaic Series). It is of coal measure facies (called the 'Coaly Grit' by William Smith in 1832), consisting on the coast of level-bedded mudstones and siltstones (25 m) on current-bedded sandstone (5 m) on a massive quartz sandstone (13 m), the last named—because it produces much of the inland moorland—the 'Moor Grit'. Filled-in channels are common in the coast sections (Figure 12). At Gristhorpe Bay the thickness approaches 40 m, increasing northwards to about 70 m; similar thicknesses are maintained across the moors but the formation is thin in the Howardian Hills and only just recognisable in the Kirkham Gorge south of High Hutton.

Dinosaur footprints have been found fairly widely in the Scalby Formation, and two species of the fresh-water bivalve *Unio* occur in abundance near Scarborough, but the main interest is in the plants. The current-bedded sandstones and micaceous shales in the level-bedded strata contain drifted leaves of Ginkgoales, pinnules of ferns, twigs and cone scales of conifers. Much material accumulated in shallow pools on the delta surface, but well-preserved plants occur also in fine silts and sands associated with the channel fillings, although logs only have survived in the coarser sandstones.

Cornbrash is probably present in the Hambleton Hills in the western part of the Cleveland Basin, as a thin sandy limestone, but to the east it is first clearly identifiable at Bilsdale. It is continuous from there eastwards to Newton Dale and Scarborough. In the east it is described as bipartite, with a black, shelly limestone, partly oolitic, above and a hard red and purple limestone beneath, totalling 0.3 to 1.5 m. The lower bed passes into shale in a thinner development on the coast. Both beds are highly fossiliferous and yield characteristic fossils of the Upper Cornbrash, including *Lopha marshii* (Plate 10), *Liostrea undosa*, *Trigonia scarburgensis* and *Microthyridina lagenalis*. Gristhorpe Cliff (Figure

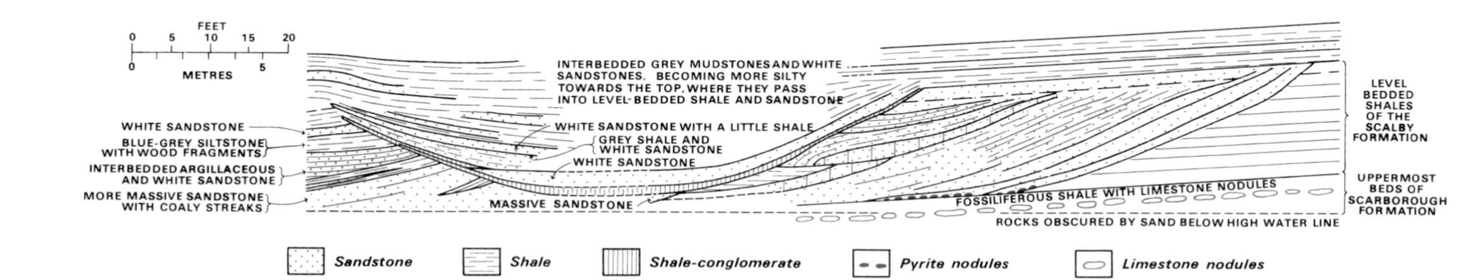

Figure 12 Channelling in the Scalby Formation on the south-eastern side of Yons Nab, Gristhorpe, Yorkshire

8), Spring Hill Brick Pit at Seamer, and Newton Dale are localities particularly rich in fossils. No representative of the Lower Cornbrash has been identified in the Cleveland Basin.

Flora of the deltaic beds

The flora of the deltaic beds in the Cleveland Basin is the world standard for the Middle Jurassic and one of the richest known. Three hundred species have been identified. The flora includes spore-plants such as *Equisetum* (horsetails) and ferns, and various groups of gymnospermous seed-plants such as conifers and cycads. The cycads and also many of the ferns are related to plants now found only in tropical and sub-tropical areas, supporting geomagnetic evidence that the region was then nearer the equator than it is now. Some of the plants belong to long-extinct groups, for example the gymnospermous group Bennettitales, whilst many others, although unlike living genera, can be placed in existing families. Only a few can be classified in living genera, for example *Marattia*, *Ginkgo* and *Equisetum*. The Ginkgo family was a particularly diverse one at the time, represented by about 15 species, one of which is shown in Figure 13. Today there is but a solitary surviving species—*Ginkgo biloba*, the Maidenhair Tree—which occurs in the wild state in a single remote region of China.

Few of the fossil plants actually grew in the places where they became fossilised, the main exception being *Equisetum* (Figure 13). Numerous rooted stems of this plant inhabited dense swamps covering large areas of ground. Most of the other plants in the flora originally grew on the drier ground of river banks or inland from the delta proper, and were transported into the swamp and river muds. Erosion and collapse of stream banks, together with natural shedding of leaves and cones, fed the delta streams with plant fragments. These were transported and usually further fragmented before they finally sank and became buried in the mud. Suitable conditions for the deposition of abundant well-preserved plant material were found only in certain parts of the delta, such as the margins of sluggish river-channels, and the richest accumulations—called plant beds—are very localised. About thirty plant beds are known, for example the classic Saltwick Formation localities at Whitby and Hayburn Wyke on the coast and the one discovered recently inland at Botton Head. In addition there are about 500 localities where only a few species occur. Thin coal seams formed by *Equisetum* swamps are common and locally occurring coals up to 40 cm thick were once worked, for example in the Cloughton Formation in Baysdale.

Variations in preservation and abundance of the plants, owing to changes in their geographical distribution during life and subsequent mode of deposition, add to the complications of identifying detached organs and to difficulties in recognising ecological and evolutionary changes. Most plant beds contain a confused mass of separated organs originating from a wide variety of plants. Records of consistent associations, the structural agreement between organs and evidence of mutual attachment are used to identify various organs attributable to the same plant, for example male and females cones, leaves and stems. From this one can consider ecological associations and any changes which may have an evolutionary significance.

The most characteristic plant horizon is at the base of the Saltwick Formation and contains numerous leaves of the pteridosperm *Pachypteris papillosa*

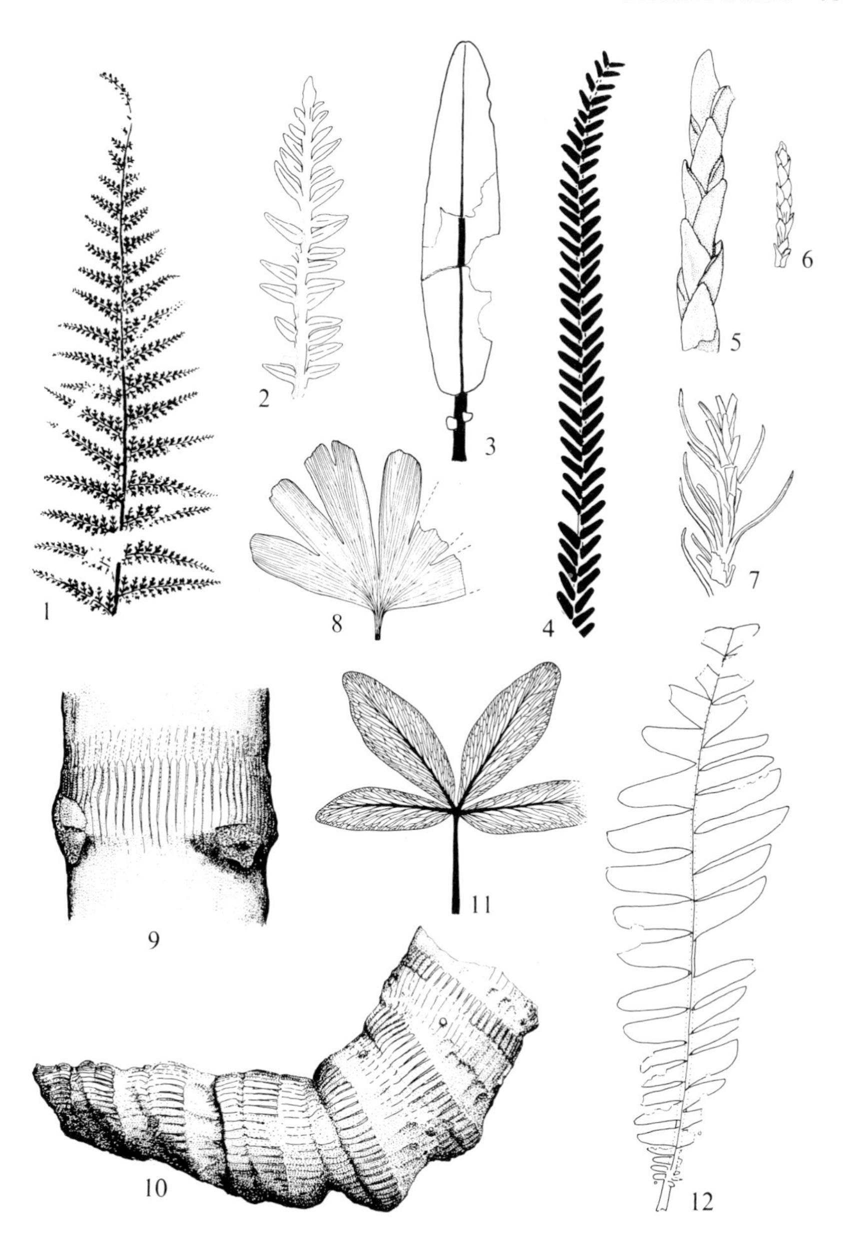

Figure 13 Middle Jurassic plants (half natural size, except 5, 6, 7 and 9)

1 *Coniopteris hymenophylloides* (Brongniart) Seward [fern]; **2** *Pachypteris papillosa* (Thomas & Bose) Harris [tree fern]; **3** *Nilssoniopteris vittata* (Brongniart) Florin; **4** *Ptilophyllum pecten* (Philips) Morris; **5, 6** *Brachyphyllum crucis* Kendall [conifer], 5×2, 6×1; **7** *Elatides williamsonii* (Brongniart) Nathorst [conifer], $\times 1$; **8** *Ginkgo huttoni* (Sternberg) Heer; **9, 10** *Equisetum columnare* Brongniart [horsetail], $9 \times \frac{2}{3}$; **11** *Sagenopteris phillipsii* (Brongniart) Presl; **12** *Nilssonia kendalliae* Harris [cycad].

(Figure 13), formerly named *Thinnfeldia*. This plant may have colonised land at the forward edge of the delta.

The particular reconstruction *Sagenopteris*+*Caytonia*+*Caytonanthus* (Figure 13), called informally the '*Caytonia*' plant, is of considerable historic interest. Scientists 50 years ago realised that the fruits of '*Caytonia*' were very like the berries of many flowering plants (angiosperms), and thought they were closed at pollination just as in an angiosperm flower. The plant thus became regarded widely as the earliest true angiosperm. Careful microscope studies, however, have shown that the fruits were open at pollination, the fertilising pollen grains actually occurring inside the entrance points to the seeds—indicating that the seeds were technically naked (gymnospermous) at fertilisation. Even thus interpreted as a highly elaborate gymnosperm, the plant retains much botanical interest. It had achieved a level of morphological complexity comparable with flowering plants even though undoubted angiosperms do not appear in the fossil record until some 70 million years later.

Western North Sea

The North Sea off Lincolnshire and Yorkshire has not provided detailed knowledge of the Middle Jurassic sequence—partly because there are few wells within 50 km of the coast, and partly because the limited sample evidence and electric logs (in the absence of cores) are difficult to interpret in terms of the detailed land stratigraphy.

It is, however, clear that the attenuation due to the Market Weighton Block continues some 40 km off the coast, and that the thick development of the Cleveland Basin is continued eastwards (Figure 10) as a predominantly sandy sequence.

In the West Sole Gasfield there is only a limited thickness of sands and shallow-water limestones, and these appear to be mainly 'Corallian'; it thus appears that the English Middle Jurassic basin ends 30 to 50 km E of the coast. Another basin with a comparable sequence of Middle Jurassic paralic sediments is known farther east, in the Dutch Offshore Graben (Heybroek, 1975). It is notable that both these basins appear to have been inverted in Upper Cretaceous times (see pp. 114–115)

7. Upper Jurassic

With the Upper Jurassic the long interval of paralic conditions came to an end in favour of fully marine deposition. Much of the sequence is argillaceous, and in consequence exposures are poor and infrequent, and knowledge is limited. The sandstones of the Callovian and the sandstones and limestones of much of the Yorkshire Oxfordian are the main exceptions, and are feature-forming rocks, and in Lincolnshire there is a minor sand development of Kimmeridgian age and a Volgian (Portlandian) sand development.

The Callovian, Oxfordian, Kimmeridgian and Portlandian stages are thus all represented in the region, although the last is known only in Lincolnshire. In general the rocks thin from both north and south towards the Market Weighton Block, where the Upper Jurassic is cut out beneath the Cretaceous for 21 km, not only at outcrop but beyond the modern coast (Figure 14). The main facies changes do not, however, coincide with this gap, which is the result of post-depositional events (see pp. 112–114) and the rocks are therefore described across the region continuously from south to north.

The earliest phase of the Upper Jurassic is represented in the region by the Upper Cornbrash, which, for convenience of description, has been included in the preceding chapter.

Kellaways Beds and Oxford Clay

From the Midlands across Lincolnshire almost to the Humber the sequence of the Kellaways Beds and Oxford Clay is relatively simple and notably uniform. The Kellaways Beds are tripartite—some 5 m of clay overlain by 5 to 8 m of sand, the top 0.5 to 1 m of which is hard and cemented and known as the Kellaways Rock. The fauna of the Kellaways Clay is not well known in this area, but the Kellaways Sand and Kellaways Rock contain common belemnites (*Cylindroteuthis puzosiana*), abundant large *Gryphaea bilobata* and occasionally the ammonite *Proplanulites*.

In Lincolnshire the outcrop of the overlying Oxford Clay is largely obscured by superficial deposits, and exposures have been infrequent. The Lower and Middle Oxford Clay are, however, beautifully exposed in the enormous brick pits from Peterborough south-westwards, beyond the limits of the region, and, in contrast to the impression of a monotonous clay sequence given by the weathered material at outcrop, are known to consist of numerous small cycles of alternating shell beds (coincident with interruptions in deposition) and mudstone. Most beds have a bituminous content and a low-grade oil-shale is present in the lowest part. The shell beds yield bivalves (especially nuculoids), and ammonites, notably *Kosmoceras*, are common throughout (Arkell, 1933; Callomon, 1968).

The Upper Oxford Clay has been cored in The Wash in connexion with the Water Storage Scheme Feasibility Study, and proves to be a more uniform sequence of pale grey slightly calcareous mudstones. The sequence appears to

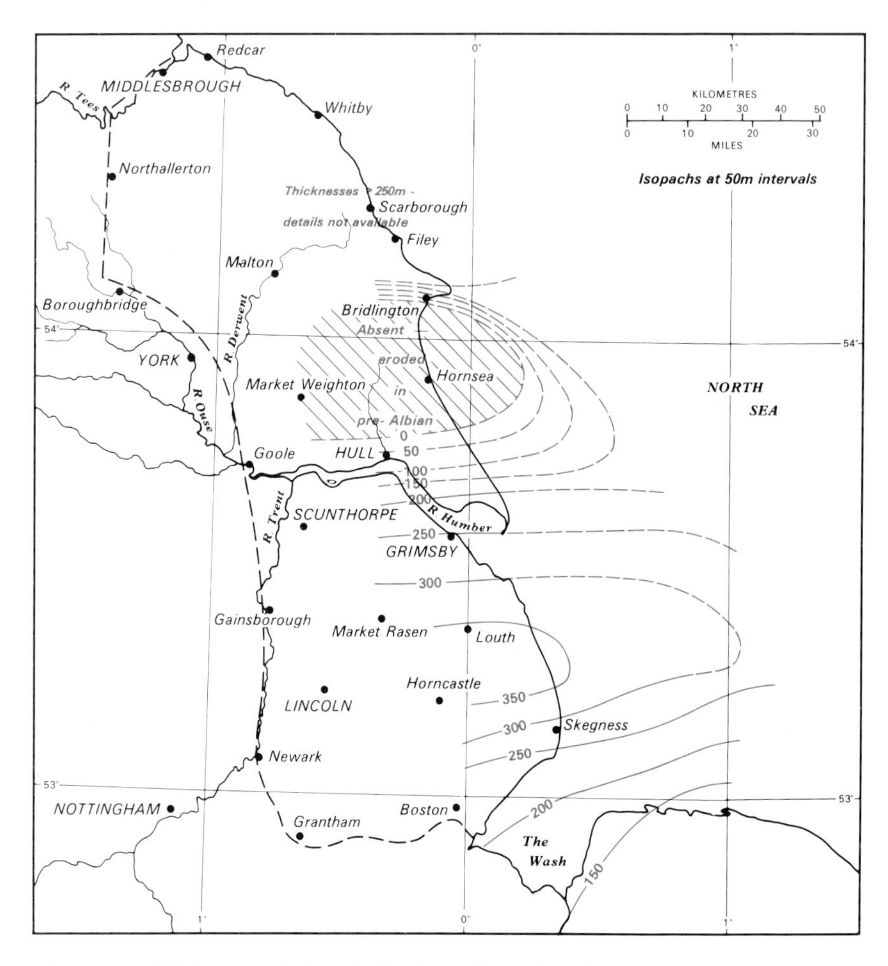

Figure 14 Thickness variations in the Upper Jurassic rocks

be much the same across Lincolnshire and is noted for beautifully preserved and locally abundant cardioceratid ammonites.

In this region the thickness of the Oxford Clay is at a maximum in the northern fenland (about 80 m), with a gradual attenuation to about 50 m near Brigg and a more marked thinning farther north. The formation is locally absent beneath a basal sandy facies of the Ampthill Clay on the Humber near North Ferriby, and disappears beneath the sub-Cretaceous unconformity near North Newbald.

The Kellaways Clay becomes discontinuous at the Humber and near South Cave is no longer present. North of the Humber there is a thickening of the Kellaways Sand, which, with the overlying Kellaways Rock, reaches a maximum of 10.5 m at South Cave. The dominant fossils are still belemnites and *Gryphaea bilobata*, but at South Cave ammonites also are common in the Kellaways Rock; these include species of *Sigaloceras* which show that the Kellaways Rock of this area is younger than that farther south and is of the same age as the basal Oxford Clay of Oxfordshire, demonstrating the beginning of northward lateral transition of clay into sandstone which continues to the Scarborough district.

North of the Market Weighton Block the Kellaways Beds and the Oxford Clay reappear from beneath the sub-Cretaceous unconformity near Garrowby. On the westerly outcrops east of Thirsk the sequence is more like that of the Midlands than that of the Yorkshire coast, with the sandy development limited to the *calloviense* Zone and consisting of soft sandstone containing concretions, quartz pebbles (in the extreme north) and the characteristic *Gryphaea bilobata* and belemnites. There are up to 35 m of Oxford Clay directly above, but this formation is locally absent, as at Roulston Scar, and measures only 13 m at Ampleforth.

The north-eastern outcrops from Newton Dale to the coast at Scarborough and Cayton Bay present a different picture (Figure 8). The Upper Cornbrash limestone is overlain on the coast by 3 m of shale, the 'Shales of the Cornbrash', which belong to the higher part of the *macrocephalus* Zone. The overlying Kellaways Rock is a chamositic sandstone, 7 to 16 m thick, with ammonites including the genera *Kepplerites* and *Proplanulites*, indicating the *calloviense* Zone. It is overlain with gentle unconformity by Middle Callovian (*coronatum* Zone) sandstone, the Langdale Beds, measuring 15 m at the type area, but only 2 to 5 m near Scarborough. Deposition was then interrupted again, and the area gently folded and subjected to erosion, so that the Upper Callovian Hackness Rock truncates both the Langdale Beds and the Kellaways Beds (Wright, 1968).

The Hackness Rock (named by William Smith after the village near Scarborough where he lived), contains ammonites of both *athleta* and *lamberti* zones (*Hecticoceras*, *Kosmoceras*, *Peltoceras* and *Quenstedtoceras*) and is thus clearly the same age as the Oxford Clay of the Midlands. It is typically a calcareous sandstone with interbedded limestones, both containing chamosite ooliths; ammonites, bivalves and gastropods are abundant. The thickness on the coast is usually under 2 m, in contrast to the tens of metres of clay of the same age in the south of England.

The Oxford Clay is only poorly exposed in the Scarborough district, but has yielded a rich ammonite fauna at Castle Hill (*Aspidoceras*, *Cardioceras*, *Creniceras*, *Quenstedtoceras*) sufficient to show that it represents only part of the *mariae* Zone.

Ampthill Clay and Corallian

Two main facies of the Middle and Upper Oxfordian are present in the region (Figure 15): the Lincolnshire development is Ampthill Clay, but the north-east Yorkshire development is principally one of shallow-water calcareous sandstones and limestones of Corallian facies. The boundary between the argillaceous and calcareous facies was originally assumed to coincide with a 'barrier' at Market Weighton, but it is now known that the Ampthill Clay extends north of Market Weighton. All the Upper Oxfordian rocks of the Acklam area are of this facies, and the clays interleave with the Corallian as far north as Malton.

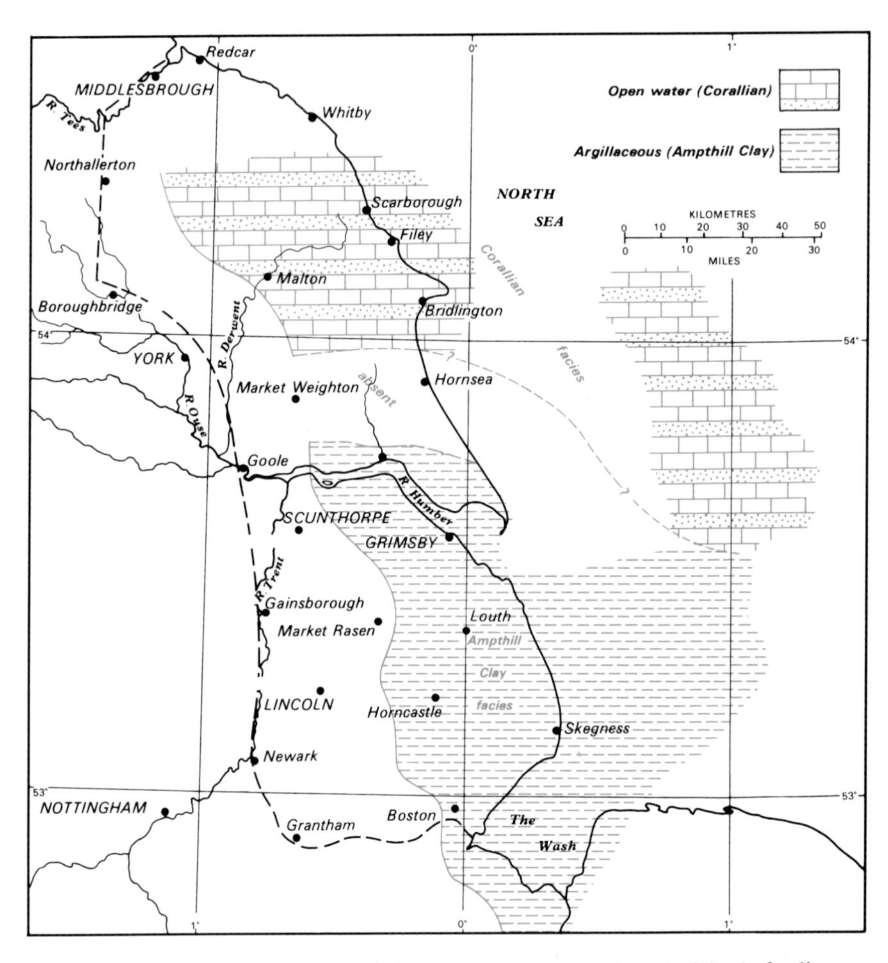

Figure 15 Geographical variation in facies during the middle part of the Oxfordian Stage

In Lincolnshire the Ampthill Clay is not dissimilar to the underlying Oxford Clay, though boreholes have shown that the mudstones are generally more silty. A characteristic feature is the occurrence of septarian nodules with masses of the worm *Serpula tetragona*, which also occur in the mudstones. The outcrop occupies approximately the middle third of the Upper Jurassic clay belt from the Fens northwards to the Humber, with a thickness estimated at about 65 m near Stixwould and Minningsby, more than 100 m at Osgodby and more than 90 m in the Humber borings. The Ampthill Clay continues eastwards at depth to the coast (Tetney Lock near Grimsby, and Winestead in Holderness), remaining entirely argillaceous with no significant intercalations of sandstone or limestone.

In the southern Wash area the Ampthill Clay rests on silty mudstones with stone bands (10 to 15 m) which are correlated with the Elsworth Rock of Cambridgeshire. There is no topographic indication of hard beds in this position at outcrop in Lincolnshire although thin limestones occur, as is indicated near Bardney, but west of Hull (Alandale, East Clough and South Cave boreholes) up to 18 m of sandstones and siltstones with thin limestones are developed at this horizon. *Gryphaea dilatata* occurs in the lower part of the Ampthill Clay and *Deltoideum delta* in the upper. Upper Oxfordian ammonites have been recorded and are characteristically non-pyritised, in contrast to those from the Oxford and Kimmeridge clays, and the formation is the presumed source of large beautifully preserved '*Amoeboceras serratum*' which are common in the boulder clay southwards into Leicestershire and Northamptonshire.

The only major exposures of the Ampthill Clay in the region are in a quarry at South Ferriby and, north of the Humber, in the Melton Clay Pit, which shows some 20 m of grey mudstones dipping eastwards at a slightly higher angle than the overlying Cretaceous. The mudstones include a bed crowded with *Serpula tetragona*, and large ammonites—*Amoeboceras* (Plate 11), *Decipia* and *Perisphinctes*. In a nearby borehole sandy limestone with the sponge *Rhaxella* is interbedded with the clays, an indication of the Elsworth Rock facies.

Brown Moor Borehole, east of Acklam and north of the Market Weighton Block, proved an incomplete section (25 m) of Ampthill Clay on 53 m of Corallian sandstones with thin limestones.

A major stratigraphical break develops at the base of the Oxfordian succession northwards from Brigg towards the Market Weighton Block, rocks of Elsworth Rock facies locally transgressing on to Kellaways Beds. Oxfordian rocks are cut out beneath the sub-Cretaceous unconformity over the Market Weighton Block. When they reappear to the north in the Acklam area the stratigraphic break is still present, for the Birdall Calcareous Grit here rests on Oxford Clay of the *athleta* Zone.

The main transition from the largely argillaceous facies of the Midlands to the arenaceous and calcareous 'Corallian' facies of north-east Yorkshire lay somewhat north of the Market Weighton Block until late Oxfordian times, during which the argillaceous facies extended northwards at least as far as the Vale of Pickering. At least a partial barrier to the flood of muddy water in the south seems to be indicated in the early and middle Oxfordian—most probably by a depth change along one of the Howardian Hills faults, with deeper water to the south, as Wright (1976) has suggested. The significance of this change is further illustrated by the offshore data and the relationship of the argillaceous beds to the limits of the East Midlands Shelf (see below, pp. 75–76).

Much of the Oxfordian succession of north-east Yorkshire is of Corallian facies—calcareous sandstones and limestones including oolites and coral reefs—forming a complex group of sediments 100 to 150 m thick. The different lithologies interdigitate with and grade into each other, but ammonites have been found sufficiently widely to establish their zonal relationships (Figure 16).

Rocks of Corallian facies extend from the *cordatum* Zone of the Lower Oxfordian to the *serratum* Zone of the Upper Oxfordian and in the Vale of Pickering and Acklam area are overlain by Ampthill Clay. The Corallian outcrop is horseshoe shaped, open to the coast and peripheral to the faulted

Vale of Pickering Syncline. The northern part, that is, the southern flank of the Cleveland Anticline, forms the Tabular Hills with strong northward-facing scarps between Scarborough and Helmsley; west of this the Hambleton Hills terminate in a west-facing scarp rising to about 400 m above the Vale of Mowbray. South of the Coxwold–Gilling Gap, in the Howardian Hills, the Corallian is thinner and the topographical expression less marked in a belt which is heavily faulted. This thinning continues southwards towards the Market Weighton Block.

The Corallian of north-east Yorkshire consists of three formations—from below upwards the Lower Calcareous Grit, the Coralline Oolite and the Upper Calcareous Grit (Wright, 1972). The 'grits' are neither true grits nor true limestones, but consist largely of fine-grained calcareous sandstones. Variations in lithology occur in both grits and oolites, but are most marked in the latter, with which reefs are also associated. Changes in the nature and profusion of the faunas are also more pronounced in the oolites.

Lower Calcareous Grit This formation provides the foundations of Filey Brigg (Plate 13) and gradually rises from below sea level to form the fine cliffs between the Brigg and Gristhorpe Bay (Figure 8, see also front cover). The outcrop extends inland through the Hackness and Tabular hills, where it is responsible for a series of abrupt north-facing scarps, as far as the western escarpment of the Hambleton Hills. Thence it swings south-eastwards, is broken by the Coxwold–Gilling Gap and continues through the Howardian Hills to the foot of the Chalk Wolds.

A gradual passage from the Oxford Clay to the Lower Calcareous Grit is apparent throughout much of north-east Yorkshire but at Roulston Scar, in the south-west part of the Hambleton Hills, the Grit lies directly on the Hackness Rock.

The Lower Calcareous Grit has its maximum development in the Tabular and Hambleton hills. It thins towards the Market Weighton Block in the south and there is also eastward thinning, particularly from the Hackness Hills to the coast, where the diminution in thickness amounts to 28 m in 10 km. The eastward thinning is associated with evidence of increasingly shallow-water deposition, and it is postulated that the formation dies out a little to the east of the present coast.

Coast sections in the Lower Calcareous Grit show the following details:

	m
3. The 'Ball Beds'; incohesive very ferruginous sandstone containing large gritty fossiliferous limestone doggers	3–6
2. Hard grey siliceous grit	1–2
1. Thickly bedded hard buff grit with a siliceous cement frequently concentrated into small masses which weather out in irregular nodular bands	13–15

These beds are recognised throughout the outcrop westwards to Rye Dale, but in the escarpments north of Kirbymoorside and Helmsley the Grit becomes intensely siliceous, owing to the abundance of the siliceous spicules of the sponge *Rhaxella perforata*, and chert bands are common. In the Howardian Hills the 'Ball Beds' are not developed; here the Grit varies from a hard siliceous spicule-bearing rock in the west to gritty limestone-centred beds and soft sandstones in the south-east.

Plate 11 Ampthill Clay and Kimmeridge Clay fossils (natural size)

1a, b *Aulacostephanus* (*A.*) *fallax* Ziegler; Kimmeridge Clay, *Aulacostephanus antissiodorensis* Zone. **2** *Rasenia evoluta* (Salfeld MS) Spath; Kimmeridge Clay, *Rasenia cymodoce* Zone. **3** *Pictonia sp.*; Kimmeridge Clay, *Pictonia baylei* Zone. **4** *Prorasenia sp.*; cementstone layer at base of Kimmeridge Clay, *Pictonia baylei* Zone. **5** *Amoeboceras sp.*; Ampthill Clay, *Amoeboceras serratum* Zone. **6** *Amoeboceras glosense* (Bigot & Brasil); Ampthill Clay, *Amoeboceras glosense* Zone.

Fossils are not common in the Lower Calcareous Grit, but fine specimens of the ammonites *Aspidoceras* and *Cardioceras* have been obtained together with the small brachiopod *Thurmannella thurmanni*, the oysters *Nanogyra nana* and *Gryphaea dilatata*, other bivalves such as *Chlamys fibrosa*, *Trigonia triquetra* and *Modiolus bipartitus*, and the echinoids *Nucleolites scutatus* and *Collyrites bicordata*.

Coralline Oolite Subdivision of this formation is shown on Figure 16. On the coast and in the Tabular Hills the two main oolite members are clearly separated by a sandstone member, the Middle Calcareous Grit, which thins out at the western end of the Vale of Pickering. In the west another sandstone member, the Birdsall Calcareous Grit, is developed in the lower part, and is in part continuous with the underlying Lower Calcareous Grit. The various members are described in upward succession.

The Birdsall Calcareous Grit is a buff sandstone which becomes increasingly calcareous upwards and is about 15 m thick. Locally it is separated from the Lower Calcareous Grit by a thin sandy or muddy oolite, the lower leaf of the Hambleton Oolite. Fossils in the easterly area include the subzonal index ammonite *Cardioceras costicardia*, *Chlamys fibrosa* and *Gryphaea*. Farther west the Grit includes the whole of the *cordatum* Subzone and extends up to the Middle Calcareous Grit.

On the coast the Lower Calcareous Grit is succeeded by the Passage Beds Member (Figure 8), which consists of calcareous sandstones, impure shelly limestones and thin-bedded sandy limestones, locally with a rich fauna of bivalves, the echinoid *Nucleolites* and occasional ammonites. In the Hackness area a 4-m coral-sponge bed, the Coral-Sponge Rag, occurs at the top of the Passage Beds. The Coral-Sponge Rag consists of an intimate mixture of masses of compound coral and sponges along with a profuse associated fauna in a matrix of reef detritus and fine mud. The most abundant fossils from the numerous quarries in this deposit include the sponges, *Enaulofungia* [*Holcospongia*] *floriceps*, *E.* [*H.*] *polita*, *Peronidella recta;* the corals *Thamnasteria concinna* (Plate 12), *Isastraea explanata*, *Rhabdophyllia phillipsi;* numerous terebratulid brachiopods, cidarid spines and many mollusca, such as *Chlamys fibrosa*, *Camptonectes lens*, *Deltoideum quadrangularis*, *Astarte ovata* and *Pseudomelania heddingtonensis* (Plate 12).

To the west of the Hackness area at Thornton Dale and Wydale the Passage Beds consist of shell-detritus limestones, at the latter place with bivalves including *Gervillia* and the ammonites *Cardioceras* and *Goliathiceras*.

The Hambleton Oolite consists mainly of oolitic limestones which in the Hackness Hills and coast sections exhibit considerable variation in lithology and fauna[1]. It is thickest around Kirbymoorside, and splits southwards into two leaves, both of which die out between Scawton and Hovingham. In the Filey district irregular hard bands with equally uneven sandy partings in the lower beds are considered to indicate intermittent intraformational erosion. The Hambleton Oolite is 9 m thick at Filey Brigg (Plate 13), but increases to about 18 m at Castle Hill, Scarborough, (Figure 8) where the upper 9 m are

[1] These beds attracted much attention from William Smith, the 'Father of English Geology', who lived at Hackness from 1828–1834.

Plate 12 Corallian fossils (natural size)

1 *Plagiostoma rigida* J. Sowerby. **2** *Pseudodiadema pseudodiadema* (Lamarck).
3 *Pseudomelania heddingtonensis* (J. Sowerby). **4** *Chlamys nattheimensis* (de Loriol).
5 *Ctenostreon sp*. **6** *Thamnasteria concinna* Goldfuss. **7** *Myophorella perlata* (Agassiz).

oolitic limestones and the rest variable sandstones and harder detrital limestones.

Certain of the beds in the Scarborough and Filey sections contain sponges, rolled fragments of *Rhabdophyllia*, cidarid spines and small terebratulids. Another noticeable feature here is the profusion of oysters and annelid incrustations, indicating shallow-water conditions.

Westwards from the Hackness area, the Hambleton Oolite thickens, becoming somewhat siliceous, often with chert centres in individual beds. These siliceous beds invariably contain spicules of *Rhaxella perforata*. Usually the oolites are poorly fossiliferous, but occasional local shelly bands, as at Wydale, Thornton Dale and Newton Dale, have yielded many fossils. One of these, the Cropton Shell Bed (Figure 16), consists of up to 4.5 m of thin-bedded impure limestones with calcareous sandstone lenses, occurring at the top of the Hambleton Oolite in a restricted area between Pickering and Kirbymoorside; its fauna includes *C. fibrosa*, *Isognomon sp.*, *Trigonia hudlestoni*, *Oxytoma expansa*, *Lima sp.*, *P. heddingtonensis* (Plate 12), *Bourguetia striata*, *N. scutatus* and *Aspidoceras sp*. Elsewhere, a thin non-oolitic limestone containing algal pisoliths is prominent near the top of the Hambleton Oolite, being traceable from Forge Valley as far west as Helmsley.

The Middle Calcareous Grit, separating the Hambleton Oolite and the Malton Oolite, is best developed between the Tabular Hills and Kirbymoorside, where it consists of up to 12 m of sandstone, generally calcareous but locally decalcified, with beds of sandy and oolitic limestones in places. Shelly layers are present, including the *Trigonia* Bed (with *T. hudlestoni*), which occurs near the top of the Grit north-west of Pickering (Figure 16). This bivalve is not uncommon in other, more fossiliferous, beds, as also are *Rhizocorallium* tubes. The best exposures are along the path approaching Newbridge Quarry and in the eastern bank of the River Doe.

Farther east, the Middle Calcareous Grit thins and becomes more calcareous, with interbedded, generally non-oolitic, limestones. It crops out on Filey Brigg (Figure 8, see also front cover and Plate 14.2), where a nodular calcareous sandstone in the upper part of the sequence contains *Cardioceras excavatum*, *Trigonia sp.* and *Gervillia aviculoides*. To the west, the Grit persists into the Helmsley area. In the Ampleforth–Gilling area a short distance to the south, the Middle Calcareous Grit rests directly on the Birdsall Calcareous Grit. In Ampleforth Quarry it consists of massive calcareous sandstone, yet less than 10 km to the east, at Ness, it has passed laterally into the lower part of the Malton Oolite.

The most persistent member of the Coralline Oolite Formation is the Malton Oolite, formerly referred to as the Osmington Oolite. It is well developed on the coast (Plate 14.2), being over 25 m thick in Cayton Carr Borehole, where the lowest 7 m are sandy, and *Thecosmilia* and *Thamnasteria* occur 9 m above the base. The Malton Oolite is traceable, generally as a thick-bedded oolite but thinning gradually and becoming sandy, as far west as Helmsley. Crossgates Quarry, Seamer, exposes 15 m, almost the complete thickness; here the lowest oolites are cross-bedded. Other quarries near Pickering and Kirbymoorside and at Helmsley provide good exposures.

Around Ampleforth the Malton Oolite is thin but locally shelly, with *Nanogyra nana*, *Gervillia sp.*, *Chlamys sp.* and echinoid and coral debris. The underlying Middle Calcareous Grit and Birdsall Calcareous Grit pass laterally into oolites in an easterly direction through Oswaldkirk and Stonegrave, and the Malton Oolite, locally more sandy, thinner bedded and less oolitic, comes to rest directly on the Hambleton Oolite in the neighbourhood of Ness. Throughout most of the Howardian Hills the Malton Oolite is predominantly oolitic but largely unfossiliferous, and is up to 30 m thick. It is well exposed in quarries at Wath, near Hovingham, and in excavations for the Malton By-pass the *Trigonia* Bed has been proved (Figure 16) within it (J. K. Wright, personal communication). South of Malton, towards Grimston, the Malton Oolite passes laterally into micritic limestone, argillaceous in the lowest part (the 'Urchin Marls'). The higher beds have yielded *N. nana*, *Trichites plotii*, *G. aviculoides*, *Chlamys nattheimensis* (Plate 12), *Lima sp.*, *Eopecten anglica*, an abundance of coral fragments and cidarid spines. Boreholes in the Acklam area still farther south suggest that the Malton Oolite is there represented by 2 m or so of interbedded silty limestones and siltstones.

The final episode of Coralline Oolite deposition is represented by the Coral Rag, a richly fossiliferous member consisting of reef limestones and detritus. The compound corals *Isastraea* and *Thamnasteria*, together with the simple forms *Thecosmilia* and *Montlivaltia*, were the main reef builders. In the interstices of the reefs occur the bivalves *Lima zonata*, *Chlamys nattheimensis*, *Lithophaga inclusa*, *Lopha gregarea* and the small gastropod '*Littorina*' *muricata*, together with the spines of *Paracidaris smithi* and *P. florigemma*.

In the Forge Valley area the Coral Rag is represented by a massive thamnasterean reef, 7 m thick, which was cut through by a channel into which reef detritus was swept and in which lived a fauna quite distinct from that inhabiting the reef. Among the fossils occurring in the channel deposit are the gastropods *Bourguetia striata* (very large and abundant), *Pseudomelania heddingtonensis* (Plate 12), *Ditremaria tornatilis*, *Nerinea visurgis*, '*Natica*' *arguta*, together with *Hemicidaris intermedia*, *Paracidaris smithi*, large numbers of the small '*Terebratula*' *kingsdownensis*, *Arca* (*Eonavicula*) *quadrisulcata* and many other bivalves.

Between Thornton Dale and Pickering the Coral Rag is attenuated in response to a local uplift which led to erosion of the underlying Malton Oolite and the formation of a boulder bed (Lee *in* Wright, 1972, p. 241).

West of Pickering the Coral Rag is characterised by the corals *Thecosmilia annularis*, *Montlivaltia dispar* and *Rhabdophyllia phillipsi*, along with the usual reef-dwelling mollusca and other fossils; it seems probable that unstable coral banks rather than true reefs existed hereabouts.

In the southern Hambleton Hills the Coral Rag extends from Oswaldkirk through the Cauklass Promontory and is identical in character with that around Sinnington and Kirbymoorside. Locally, between Helmsley and Sproxton, it is represented by a fine-grained argillaceous limestone containing stray reef fossils and occasional lumps of dark chert.

In the Howardian Hills the Coral Rag is well developed around Gilling, Hovingham, Slingsby, Malton and North Grimston; at the last-named locality it contains much nodular chert.

At Hildenley, near Malton, there is a faulted mass of pure white limestone (Hildenley Limestone) with a few reef fossils; this represents fine calcareous mud washed from some nearby reef no longer visible. The biomicritic facies is

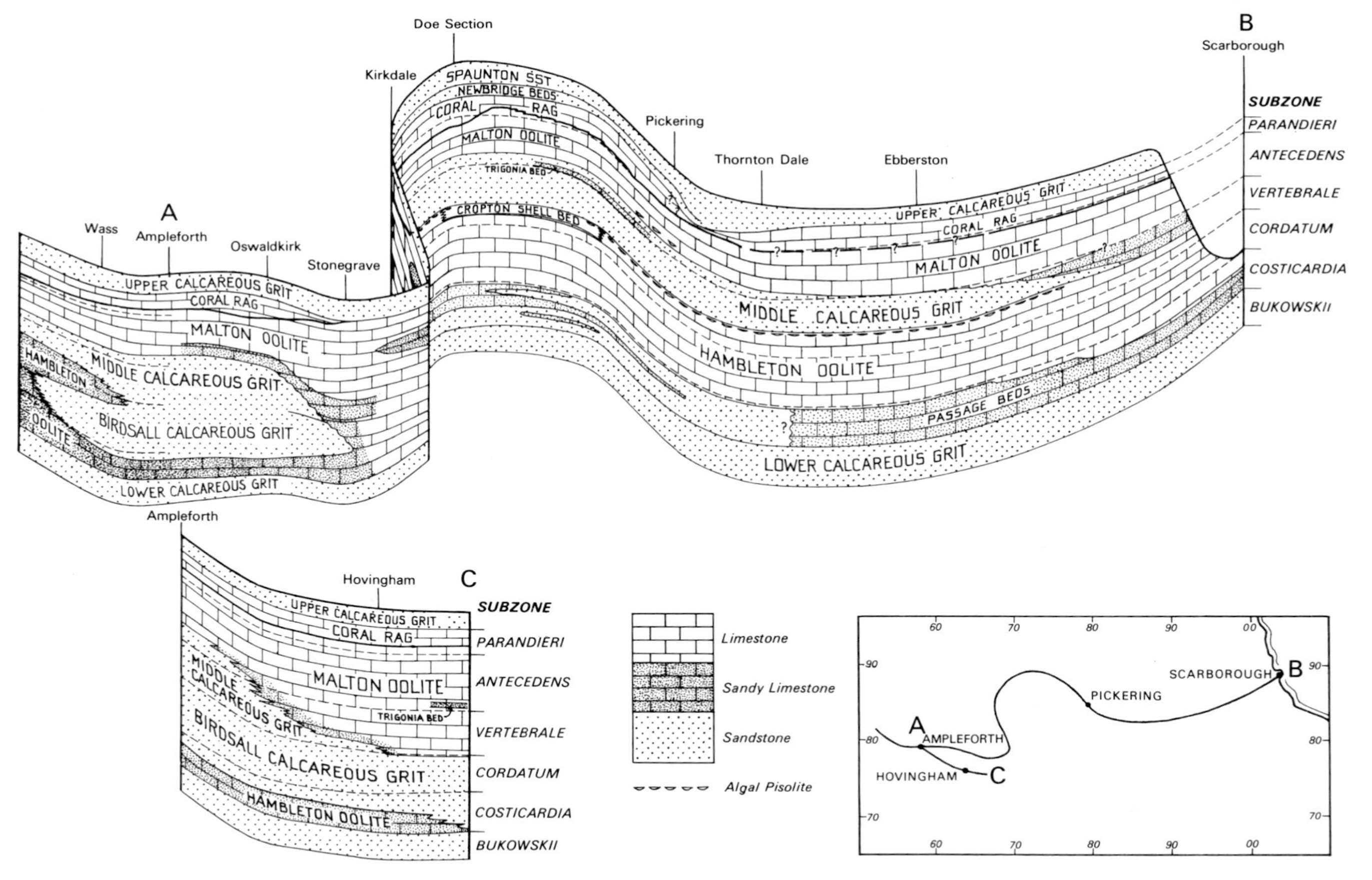

Figure 16 Ribbon section of the Corallian rocks of Yorkshire

transitional to that of the overlying North Grimston Cementstones. At the close of Coralline Oolite sedimentation conditions become inimical to coral growth. Submarine erosion levelled off the banks and reefs of the Coral Rag and the detritus, along with finer incoming arenaceous material, was deposited as Upper Calcareous Grit.

Upper Calcareous Grit The outcrop of this formation forms an intermittent cap to some of the southern ridges of the Tabular Hills, but is continuous from Pickering westwards through Helmsley into the south-eastern part of the Hambleton Hills and the Cauklass Promontory. Throughout the whole of this area it is largely a fine-grained sandstone up to about 15 m thick. In it, the bivalves *Gryphaea dilatata*, *'Lucina' fulva*, *Chlamys fibrosa*, *C. midas* and *Camptonectes lens* and fragments of ammonites are fairly common. There are excellent sections in these beds at Pickering, in Hutton Beck and at Nunnington. Ammonites are particularly abundant in a quarry at Nunnington; these prove to be a distinctive fauna (known also in Greenland) in which the index fossil is *Amoeboceras nunningtonense* (Wright, 1972).

At Snape Hill, near Kilburn, the Upper Calcareous Grit is represented by about 10 m of alternating hard and soft beds of fine calcareous mudstone overlain by 0.5 m of limestone and at least 8 m of Snape Sandstone. Similar beds occur farther to the south-east between Langton and North Grimston, where they are known as the 'North Grimston Cementstones'; about 12 m are visible, but the full thickness is estimated at about 17 m. They have yielded large numbers of the oyster *Gryphaea dilatata*, together with *Gervillia sulcata*, *Pholadomya hemicardia*, *Nanogyra nana* and fragments of belemnites and ammonites.

Formerly it was thought that the Kimmeridge Clay lay unconformably on the North Grimston Cementstones, but it is now known that there is a gradual passage from the latter up into Ampthill Clay referable to the higher zones of the *Oxfordian*.

Offshore

Well sections in the south-western North Sea provide particularly significant information on the Middle and Upper Oxfordian rocks, despite the fact that the dating is generalised—assignment of the beds to individual zones is not yet possible—and that the later Jurassic sediments are missing from a broad belt east of Holderness (Figure 14).

The Corallian facies has been proved near the coast in the Scarborough area (Dingle, 1971), and it is known from there south-eastwards across the West Sole Gasfield—a distance of at least 100 km, extending down to the latitude of the Humber. In this area the facies is dominantly one of limestones, commonly pisolitic, with subordinate sandstones.

From the latitude of Grimsby southwards the succession is argillaceous (Ampthill Clay facies) as in Lincolnshire, with sandy intercalations in the wells nearer the coast and interbedded thin limestones in an area 50 km offshore which suggest the continuation of the Elsworth Rock facies of The Wash.

Although these data leave a large gap south-south-east of Flamborough Head, there is a broad coincidence of the argillaceous deeper water facies of the Middle/Upper Oxfordian with the East Midlands Shelf, and of the shallow-water Corallian with the Cleveland Basin and the basinal area east of the shelf offshore. This is at variance with the general buoyancy of the East Midlands Shelf and of the Market Weighton area in particular through Jurassic time, and indicates that inversion of the basinal area was already beginning (see Chapter 11).

Kimmeridge Clay

The very scanty stratigraphical information previously available on the Kimmeridge Clay has recently been augmented by The Wash Water Storage Scheme Feasibility Study (Gallois, 1974) and by an investigation of oil shale potential in Lincolnshire (Figure 17).

The sequence in southernmost Lincolnshire is in general argillaceous. In detail it comprises a complex of small-scale rhythms each 0.3 to 2.5 m thick, consisting of soft mudstones, shelly mudstones, calcareous mudstones and relatively rare thin argillaceous limestones or lines of cementstone nodules. Oil shales of varying richness occur at many horizons. Most of them are thin and unimportant, but there are 5 m of bituminous beds at the top of the *eudoxus* Zone and at numerous higher horizons.

The Kimmeridge Clay as seen in the cored boreholes in The Wash area differs from the Ampthill Clay in its greater range of lithologies, in the common occurrence of oil shale and smooth dark mudstones, and in the general absence of pyrite. The maximum thickness recorded in The Wash area (130 m) continues into mid-Lincolnshire and oil shale horizons continue to be numerous, as originally found in a boring at Donington-on-Bain, probably drilled in 1815. The Kimmeridge Clay was formerly exposed in many small brickyards west of the Lincolnshire Wolds, and these yielded a rich fauna. The basal beds (formerly worked at Stickney, near Boston) are blue clays which contain the smooth ammonite *Pictonia* (Plate 11) in abundance; somewhat higher beds contain *Rasenia*. Market Rasen was the type locality of the latter, yielding in the last century the beautifully preserved specimens to be found in many museums. Exposures of fissile bituminous shale with abundant ammonites in the middle and upper parts are common at the foot of the Wolds scarp, particularly just below the Spilsby Sandstone. The ammonites are mostly *Pectinatites spp*., indicating the higher oil shales.

In north Lincolnshire a notable variant from the continuous argillaceous facies is provided by a medium to coarse sandstone lens in the lower part of the Kimmeridge Clay at Elsham, east of Brigg (Figure 18). This bed was formerly misidentified as Spilsby Sandstone and mapped as a detached piece of Lower Cretaceous beneath the Upper Cretaceous unconformity, but it has yielded abundant ammonites of the sub-genus *Xenostephanus*, establishing an early Kimmeridgian age.

The Kimmeridge Clay is progressively overstepped by Cretaceous rocks (Carstone and Red Chalk) north of Caistor and disappears at the Humber, to reappear in the Vale of Pickering. It is heavily blanketed by Boulder Clay and is relatively little known in Yorkshire, despite its great thickness (possibly up to 273 m in Fordon No. 1 Borehole). An up-to-date review of the zonal stratigraphy has been provided by Cope (1974).

Plate 13 Corallian rocks on foreshore at Filey Brigg [TA 130 815]. The Lower Calcareous Grit forms the lowest scar (extreme left) and is overlain by the Passage Beds and the Hambleton Oolite. The Middle Calcareous Grit forms the highest scar (right of centre), with till in right foreground. (L 1344)

Plate 14 1 Deltaic sandstones in Saltwick Formation, Snargate Hill Quarry [SE 608 722], near Brandsby in the Howardian Hills. (L 1682)

2 Casts of *Thalassinoides* burrows on displaced block of Middle Calcareous Grit, Filey Brigg [TA 131 815]. (L 1345)

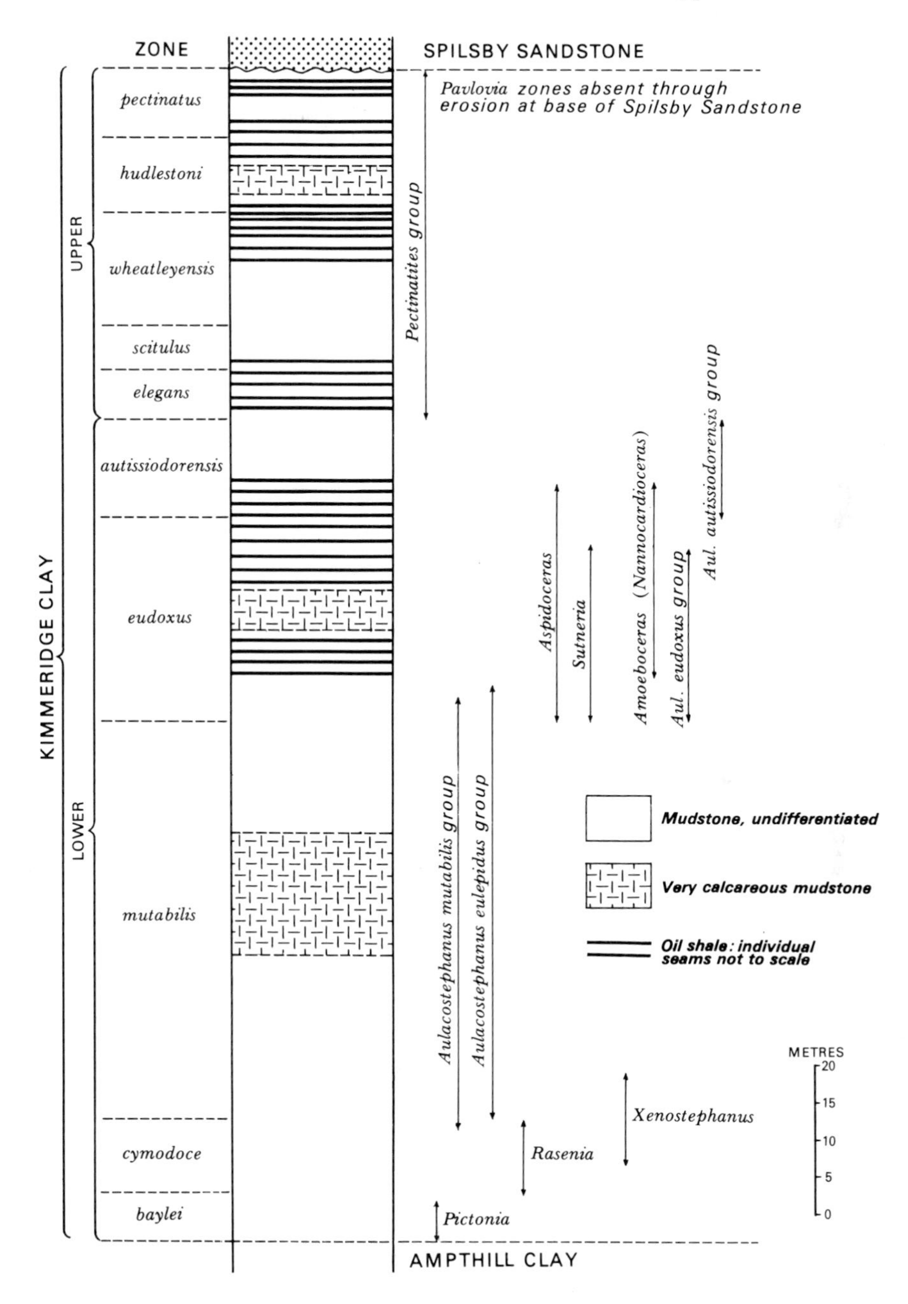

Figure 17 Generalised section of the Kimmeridge Clay of Lincolnshire

The lithology of the Yorkshire succession, as far as it is known from borings, is much the same as in Lincolnshire, though oil shales appear to be less significant. The exposures of the lower zones that have yielded rich faunas in the

Malton area are now no longer available. The *eudoxus* Zone is still well exposed at Marton, west of Pickering, and higher beds are seen in the nearby Golden Hill Pit, where there is a discontinuity above the *wheatleyensis* Zone. The *pectinatus* Zone, the highest Jurassic zone seen in Yorkshire, is present here, as it is on the coast near Speeton. Higher zones (*rotunda* and *pallasioides*) of the Kimmeridgian, and equivalents of the Volgian (Portlandian), were either not deposited or were eroded before the accumulation of the Lower Cretaceous rocks.

Lower Spilsby Sandstone

In this region post-Kimmeridgian Jurassic strata are limited to Lincolnshire, where the Lower Spilsby Sandstone, until recently regarded as Lower Cretaceous (Neocomian, Infra-Valanginian), is now known on ammonite evidence to belong to the Volgian Stage of Russia, equivalent in age to the Portland and Lower Purbeck beds of southern England (Casey, 1962). *Subcraspedites*, which occurs in the basal Spilsby Sandstone at Partney, near Spilsby, was formerly believed to be a Cretaceous genus, but is now known to be Jurassic; the associated *Paracraspedites* is probably syngeneric with *Titanites* from Dorset (the former commonly lacks the outer whorls, the latter the inner stages), and the Portlandian genera *Kerberites* and *Crendonites* have been found in the basal Spilsby Sandstone at Nettleton, near Caistor. The mid-Spilsby nodule bed is of late Ryazanian age, but there is a non-sequence below it, and its base is taken as the boundary between the Jurassic and Cretaceous systems in Lincolnshire. For convenience the Spilsby Sandstone (Figure 18) is described as a single lithological unit in the next Chapter.

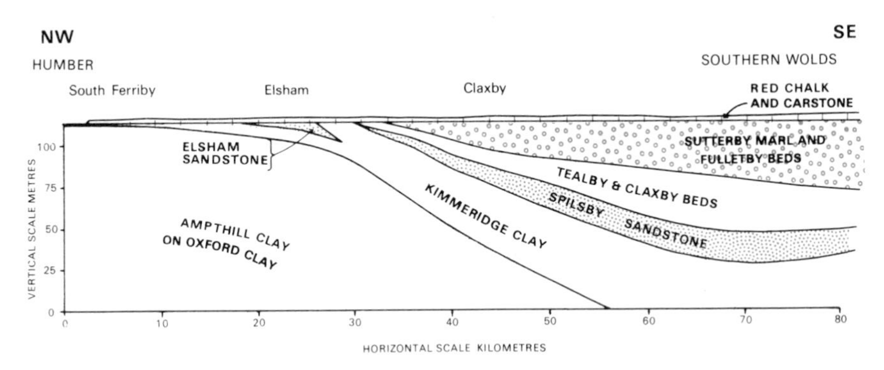

Figure 18 Generalised section of the younger Upper Jurassic rocks and the Lower Cretaceous rocks across Lincolnshire

Offshore

The nearer North Sea wells give a strongly varying picture of Kimmeridgian deposition. The full succession appears to be linked to the area south of the Humber as far out as 1°E, where there is one record of 235 m and two others of a little over 150 m. In the area where the Corallian facies is developed in the underlying Oxfordian rocks, however, the sequence is incomplete, with all the

Upper Kimmeridge Clay and parts of the Lower Kimmeridge Clay cut out at the sub-Cretaceous unconformity. This is an effect independent of the bouyancy of the Market Weighton Block (although the area lies largely in line with it), being coincident with thick developments of older Jurassic strata, and is to be ascribed to continued inversion of the Cleveland–Sole Pit basin.

8. Lower Cretaceous

Although in north-eastern England sedimentation in the latest Jurassic and early Cretaceous was discontinuous, there were no major periods of non-marine deposition comparable to the Purbeck and Wealden of southern England. The Neocomian stages—Ryazanian, Valanginian, Hauterivian and Barremian—are all developed and are represented by richly fossiliferous marine beds. Also in contrast to southern England, the Aptian and Albian strata are very thin (less than 17 m) and form only a minor part of the Lower Cretaceous succession.

Neocomian and Aptian

In Lincolnshire Neocomian and Aptian rocks are present through the southern two-thirds of the Wolds. They are overstepped by Albian Carstone and Red Chalk near Caistor without major change of facies, but reappear in east Yorkshire, where they underlie only the north-easternmost corner of the Chalk Wolds eastwards from Heslerton. At Heslerton they are already developed to full thickness, which continues to the Fordon area, thinning from there towards the coast. In Lincolnshire the sequence is lithologically very varied—sandstones, limestones, ironstones and clays; in Yorkshire it is developed as clay only (the Speeton Clay), which covers a similar time-range (Table 5).

The most complete succession is located in the southern Lincolnshire Wolds (Figure 18). The Jurassic clays are there overlain by the Spilsby Sandstone, up to 25 m of fine- to medium-grained sandstone, strongly glauconitic in the lower part, largely uncemented but with calcareous cemented masses varying in size from 1 to 3 m. Fossils are mainly preserved in these calcareous concretions. Phosphatic nodule beds with both derived and indigenous fossils are developed at two horizons—the basal layer contains derived Kimmeridgian ammonites (*Pavlovia sp.*) and a contemporaneous Upper Jurassic (Volgian) fauna, including *Kerberites* and *Crendonites* and the massive belemnites *Acroteuthis partneyi* and *A. lindseyensis* (Plate 15). The mid-Spilsby nodule bed is taken as the base of the Cretaceous, corresponding to the middle Purbeck Cinder Bed of southern England; it contains ammonites including the indigenous Ryazanian *Surites* and the derived Jurassic (Volgian) *Subcraspedites*. No representative of these Lower Ryazanian beds is known in Yorkshire.

The Spilsby Sandstone is overlain by the Claxby Ironstone, a brown to purple oolitic ironstone up to 5 m thick, which has been worked at Acre House and Claxby, near Caistor, as an iron ore for admixture with Liassic ores. This bed becomes shaly south-eastwards—comprising the deeper water Hundleby Clay facies—but limonite ironstone intercalations are still present as far south as Spilsby, near the centre of the original basin. The Claxby Ironstone contains a particularly rich fauna of mollusca, including numerous belemnites and bivalves, with less common ammonites. The ammonites include *Endemoceras* [*Lyticoceras*] *oxygonium*, *Polyptychites sp.* and tolliids; the belemnite

Plate 15 Spilsby Sandstone and Claxby Ironstone fossils (natural size, except 4 and 6)

1 *Peregrinoceras albidum* Casey; Upper Spilsby Sandstone (Cretaceous). **2, 3** *Rouillieria tilbyensis* (Davidson); Claxby Ironstone. **4** *Camptonectes* (*Boreionectes*) *cinctus* (J. Sowerby), $\times \frac{1}{2}$; Claxby Ironstone. **5** *Acroteuthis lindseyensis* Swinnerton; Lower Spilsby Sandstone (Jurassic). **6** *Subcraspedites* (*S.*) *sowerbyi* Spath, $\times \frac{1}{2}$; Lower Spilsby Sandstone (Jurassic).

Acroteuthis subquadratus with allied species is characteristic, and one of the most notable features is the occurrence of well-preserved specimens of the large pecten, *Camptonectes* [*Boreionectes*] *cinctus* (Plate 15).

Above the Claxby Ironstone (or above the Hundleby Clay in the south) are the Tealby Beds, some 20 m thick and consisting of the Lower and Upper Tealby Clays separated by the Tealby Limestone. These beds constitute the nearest approach in the Lincolnshire succession to the Speeton Clay facies of Yorkshire.

The Lower Tealby Clay is a dark plastic clay, often richly glauconitic, with the ammonites *Aegocrioceras*, *Endemoceras* and *Simbirskites*, the belemnite *Hibolites jaculoides* and again the large pecten *Camptonectes cintus*. The Tealby Limestone is 10 to 40 m of argillaceous or sandy yellow-weathering limestone with clay partings and scattered fossils (see Plate 16). The Upper Tealby Clay is greenish when unweathered, but becomes light coloured and ferruginous above, grading into the overlying Fulletby Beds. Its fauna includes the belemnites *Aulacoteuthis* and *Oxyteuthis* and sporadic crioceratid ammonites.

The succeeding Fulletby Beds, 20 m thick, show a reversion to iron deposition. The facies is dominantly clayey, variably sandy and ferruginous with abundant dark brown or black polished limonite oöliths. The iron content reaches a maximum in a central cemented member, the Roach Stone, which is a ferruginous sandstone some 4 to 5 m thick and locally a low-grade iron ore, although never exploited. The fauna of the Fulletby Beds includes *Mulletia mulleti*, some rhynchonellids, and belemnites dominated by *Oxyteuthis brunsvicensis*.

In southern Lincolnshire the Fulletby Beds are succeeded by the Skegness Clay and overlying Sutterby Marl, both up to about 2 m thick. The former contains Lower Aptian ammonites including *Prodeshayesites;* the latter contains the Upper Aptian belemnite *Neohibolites ewaldi* with derived Lower Aptian ammonites in a basal nodule bed.

The Yorkshire succession is adequately exposed only at Speeton, and there is no comparable evidence of lateral variation. Deposition began there rather later in the Ryazanian Stage than in Lincolnshire, so that at Speeton there is a greater stratigraphical gap at the base between the lowest horizon—the phosphatic Coprolite Bed—and the *pectinatus* Zone which comprises the highest surviving Kimmeridgian, the equivalents of the Volgian Lower Spilsby Sandstone being absent.

The sequence is almost entirely of clays, with nodule beds which provide markers. The cliffs are unstable and are heavily slipped, and the sequence (Table 5) has been compiled largely from temporary exposures between tide marks. It has been classified by the belemnite faunas.

The basal Coprolite Bed is a thin band of phosphatic nodules and contains eroded casts of Kimmeridgian ammonites. The *Acroteuthis lateralis* Beds (D6–D8) show evidence of slow accumulation which, at times, was interrupted by current action; in these beds fossils are poor, but the belemnite *Acroteuthis* is common. The *A. subquadratus* and *H. jaculoides* beds (D5–C1) are mainly banded clays which accumulated slowly in fairly deep water, and in them ammonites are abundant and enable many of the Continental Neocomian zones to be recognised. *Polyptychites* (Plate 17) and tolliid ammonites occur in the lower part, *Endemoceras* and *Simbirskites* (Plate 18) higher up. In beds D4

Plate 16 Claxby Ironstone and Tealby Limestone fossils (natural size, except 7)

1, 2 *Cucullaea* (*Dicranodonta*) *benniworthensis* Kelly; Claxby Ironstone. **3, 4** *Aetostreon subsinuata couloni* (Defrance); Tealby Limestone. **5, 6** *Lamellaerhynchia rostriformis* (Roemer); Claxby Ironstone. **7** *Simbirskites* (*Craspedodiscus*) *discofalcatus* (Lahusen), $\times \frac{1}{2}$; Tealby Limestone.

Plate 17 Speeton Clay fossils (natural size)

1, 2 *Polyptychites gravesiformis* Pavlov; D. Beds. **3** *Oxytoma cornuelianum* (d.Orbigny); C. Beds. **4, 5, 6** *Buchia lamplughi* Pavlov; C. Beds. **7** *Aegocrioceras quadratum* (Crick); C. Beds. **8** *Thracia phillipsii* Roemer; C. Beds. **9** *Meyeria ornata* (Phillips); C. Beds. **10** *Astarte senecta* Woods; C. Beds. **11** *Panopea neocomiensis* Leymerie; B/C/D. Beds.

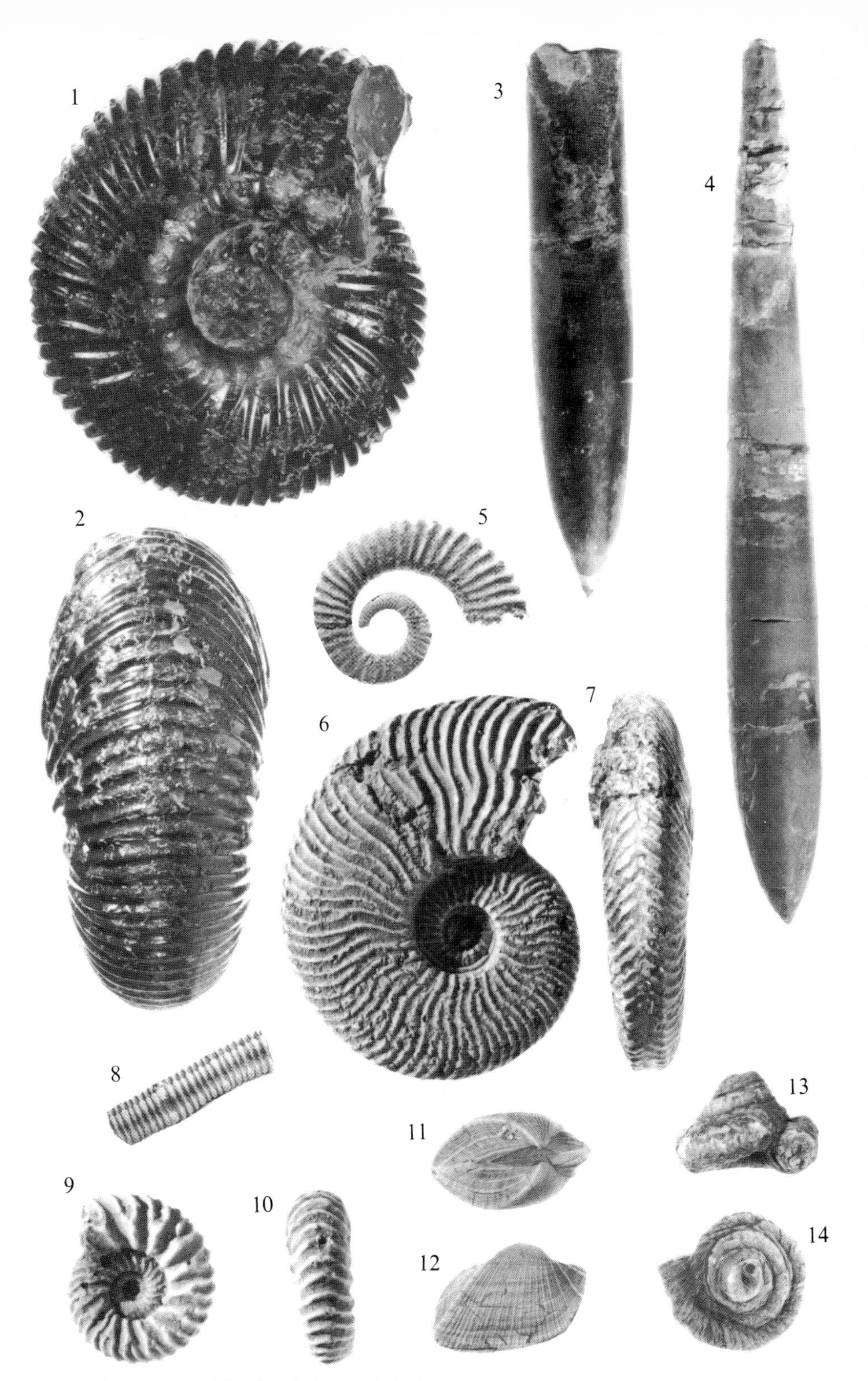

Plate 18 Speeton Clay fossils (natural size)

1, 2 *Dichotomites typicus* Spath; base of bed D2. **3** *Oxyteuthis brunsvicensis* (Strombeck); B. Beds. **4** *Hibolites jaculoides* Swinnerton; C. Beds. **5** *Aegocrioceras bicarinatum* (Young & Bird); C. Beds. **6, 7** *Endemoceras regale* (Pavlov); lower C. Beds. **8** *Isocrinus annulatus* (Roemer). **9, 10** *Simbirskites* (*Speetoniceras*) *subbipliciforme* (Spath); bed C9. **11, 12** *Grammatodon* (*Indogrammatodon*) *securis* (Leymerie); uppermost B. Beds. **13, 14** *Rotularia* (*R.*) *phillipsii* (Roemer); B. Beds.

and D5 it is interesting to note the presence of *Lingula subovalis*, a small brachiopod of a genus that has persisted from Ordovician times to the present. Bed C4 contains a remarkable line of nodules, most of which enclose the remains of a shrimp, *Meyeria ornata* (Plate 17), while from bed C3 above comes a profusion of the small echinoid *Toxaster*. The '*brunsvicensis*' Beds (B) comprise a considerable thickness of light and dark clays, and in the upper part there are numerous layers of cementstone nodules; the belemnite fauna of these beds is quite distinct from that of the *H. jaculoides* beds below. The topmost B beds contain a Lower Aptian fauna with *Prodeshayesites* and the belemnite *Hibolites minutus*. Marls and clays complete the remainder of the formation up to the Red Chalk; the lowest part, the Aptian/Lower Albian *ewaldi* Marl, is characterised by the presence of *Neohibolites ewaldi*, while *N. minimus* occurs in the higher beds—the *minimus* Marls of middle and possibly late Albian age.

Boreholes through the Chalk west of Speeton have shown the Speeton Clay to be developed to much greater thicknesses (Neale, 1974) and it is believed that distribution and thickness are both dependent on a phase of pre-Albian faulting and erosion. These movements were contemporary with the tilting and truncation of the Neocomian and older rocks from Market Weighton southwards.

Table 5 Classification of the Lower Cretaceous rocks of the region

Speeton area				**Stage**	South Lincolnshire
	Red Chalk			**Albian**	Red Chalk
Speeton Clay	*minimus* Marls (A1-3)				
	Greensand Streak (A4)				Carstone
	ewaldi Marl (A5)			**Aptian**	
					Sutterby Marl
	'*brunsvicensis*' Beds	Upper B Beds			Skegness Clay
				Barremian	Fulletby Beds
		Middle B (Cement) Beds			Upper Tealby Clay
		Lower B Beds	LB5D / LB5E		Tealby Limestone
				Hauterivian	
	A. sub-quadratus and *H. jaculoides* Beds	C Beds (1-11)			Lower Tealby Clay
		D Beds (1-5)	D2D / D2E		Claxby Ironstone
				Valanginian	Hundleby Clay
	A. lateralis Beds	D Beds (6-8)		**Ryazanian**	Spilsby Sandstone (upper part)
	Coprolite Bed (E)				

Albian

In north Yorkshire the Albian Stage is represented by the upper part of the *ewaldi* Marl and by the *minimus* Marls of the Speeton Clay (see above) with the overlying Red Chalk. Elsewhere in the region only Red Chalk with discontinuous underlying Carstone is present.

The Carstone shown on the old maps of the Lincolnshire Wolds is a sandstone, generally fine-grained below and coarse and gritty above. There are indications of a stratigraphic break in the middle part, and only the coarse gritty part is now regarded as Carstone. This bed, with the overlying Red Chalk into which it passes, transgresses across the underlying Lower Cretaceous rocks northwards and comes to rest on Jurassic rocks north of Caistor.

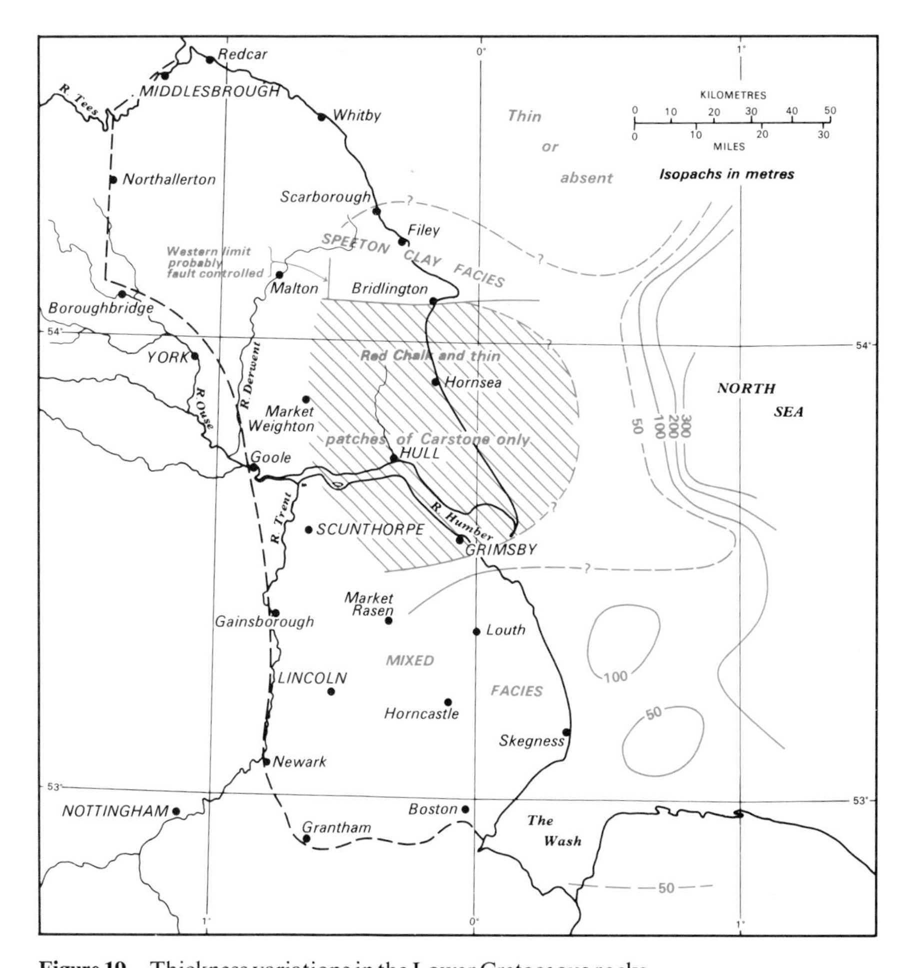

Figure 19 Thickness variations in the Lower Cretaceous rocks

Speeton Clay and mixed facies apply to beds below the Red Chalk and Carstone.

The Red Chalk is an impure limestone, varying in colour from pink to brick-red, containing rounded quartz grains, numerous fossils (see Plate 21), and grading downwards into the Carstone. It represents a condensed deposit which accumulated slowly. Its colour may be due to red mud washed from a low-lying lateritised contemporary land area, or alternatively it has been suggested that the ferruginous material could have been derived from exposed Keuper Marl on rising North Sea salt intrusions (Kent, 1967). It is continuous throughout Lincolnshire and Yorkshire, being 5.5 m thick at the southern extremity of the Lincolnshire Wolds but gradually diminishing in thickness northwards towards the Market Weighton Block, until only 1 to 2 m are found north of Clixby and in the southern Yorkshire Wolds. It is thicker again, up to 12.8 m (Jeans, 1973) farther north in Yorkshire, where the overlying basal Cenomanian Chalk is also red.

North of Caistor, the Albian rocks rapidly overlap all the older Lower Cretaceous formations, until beyond Clixby they rest directly on the eroded surface of the Upper Jurassic clays. On the northern side of the Humber they rest directly on the Ampthill Clay, then transgress across the Kellaways Beds northwards beyond South Cave and rest on Lower Lias east and north of Market Weighton—the maximum development of the pre-Albian unconformity in the region.

In the course of the northward transgression of the Red Chalk in Lincolnshire the associated Carstone diminishes in thickness and becomes progressively coarser. Between Tealby and Clixby it is exceptionally coarse and contains numerous phosphatic nodules and fragmentary casts of ammonites. Where present, the Carstone maintains a gradual transition into the overlying Red Chalk. In northern Lincolnshire it is missing locally, but the lower part of the Red Chalk is conglomeratic, as for example at Worlaby, with quartz grains and phosphatic nodules in fair abundance.

Farther north in Yorkshire, at three localities, Goodmanham, Millington and Kirby Underdale, the Carstone reappears as a metre or so of ferruginous sands resting unconformably on Jurassic strata and again passing up gradually into the Red Chalk, which is about 1.5 m thick. There are variegated ferruginous sands near Goodmanham and in Millington Dale, north-east of Millington. At Great Givendale a band of conglomerate intervenes between them and the Red Chalk. At the head of Scottendale, east of Kirby Underdale, the following sequence is exposed:

		m
Red Chalk –	Nodular red chalk	0.9
	Soft marly red chalk	0.3
Carstone –	Brown marly sand	0.3
	Coarse ferruginous sands with thin conglomerate at base	6.0
	Soft yellow micaceous sands of unknown age . . . seen for	0.6
Upper Lias–Clay		

Here the Carstone is a feebly cemented mass of coarse sand grains coated with iron oxide, along with pebbles of white quartz, black lydian stone, oolitic ironstone, buff sandstone and phosphatic nodules, and is riddled with thin veins of ironstone. The layer of conglomerate at its base is made up of bored phosphatic nodules of varying sizes in a coarser greensand matrix. The passage into the overlying Red Chalk is gradual.

North of Market Weighton the various higher Jurassic beds reappear below the unconformity, and the Red Chalk is again locally conglomeratic. Along the northern edge of the Wolds to the sea the Red Chalk rests directly on the Speeton Clay.

The Lower Cretaceous extends widely beneath the North Sea (Figure 19), as clays comparable in facies and time-range to the Speeton Clay. In this context the Lincolnshire and Norfolk development is seen as an abnormal marginal facies. On the seaward part of the East Midlands Shelf thickness changes are gentle, as in Lincolnshire, and the Lower Cretaceous follows the Kimmeridge Clay uniformly. On the edge of the shelf there is a sharp increase in thickness with a north–south trend, which coincides with the edge of the unstable salt basin, subject to strong halokinetic disturbance at this time. The Lower Cretaceous still rests on Kimmeridge Clay around 1°E, but from 2°E to the median line becomes strongly transgressive, resting on Lias 100 km E of Scarborough and on Trias between there and the Indefatigable Gasfield. In contrast to the Lincolnshire development, but in line with relationships on the East Anglian Block, the Aptian and perhaps the Lower Barremian take part in the main transgression.

9. Upper Cretaceous

Towards the close of early Cretaceous times a major eustatic rise in sea level resulted in a global transgression that progressively inundated the whole of the European area. As sea level rose, less and less terrigenous material was supplied by the shrinking land masses, and calcareous deposits which were to become the Chalk were formed. In general, the Chalk is a very fine-grained, extremely pure (more than 98 per cent $CaCO_3$), relatively soft, white limestone. Although it was previously thought to be either a deep-sea deposit like the modern Globigerina Ooze, or a shallow-water inorganic aragonite precipitate of Bahama Banks type, electron microscope studies have demonstrated that up to 90 per cent of the Chalk is composed of tiny calcite crystals a few microns across, derived from the disintegration of complex ring-like structures known as coccoliths. These are secreted by the Coccolithophoridae, a family of highly specialised unicellular green algae. Within this coccolith 'flour' are suspended larger particles of biogenic calcite, including spherical bodies of uncertain affinity called calcispheres (*Oligostegina*), foraminifera, ostracod valves, bryozoa, miscellaneous shell and echinoderm debris and, in particular, prisms derived from the mechanical breakdown of shells of the bivalve *Inoceramus*. In current-winnowed horizons the calcite prisms reach rock-building proportions and produce chalks which are gritty to the touch. Macrofossils constitute a relatively insignificant proportion by volume of the total sediment, and tend to be concentrated at particular horizons.

Also present in the Chalk are sheets and irregular nodules of flint. Flint is a chert with a particularly well developed conchoidal fracture composed of an aggregate of tiny quartz crystals of the same order of size as the coccolith plates. The sheets and nodules are predominantly aligned parallel to the bedding but are not necessarily syngenetic, whereas high-angle sheets cutting the bedding are clearly of later origin. It can be shown that the majority of flints are intimately related to trace fossils, and are replacements of the sedimentary fill of burrows or, less commonly, are formed around burrows. An extreme case of the latter is the paramoudra, a vertical flint up to several metres long, which has formed around a tiny burrow. It is now thought that the formation of flint was a multi-stage process completed before the end of the Cretaceous. Dissolved silica derived from skeletons of siliceous sponges and radiolaria was initially absorbed by organic matter concentrated in the sedimentary fills of burrows. With greater depth of burial, decomposition of the organic matter released free silica in the form of minute spheres of cristobalite, which provided nuclei for the accretion of more silica from solution. These embryonic flints composed of cristobalite and other silica minerals eventually underwent late-stage alteration to quartz.

The Chalk of the region belongs to an ill-defined lithofacial and faunal 'Northern Province' which extends southwards across The Wash into north Norfolk. In marked contrast to the massive 'earthy' Chalk of the complementary 'Southern Province', the northern Chalk is for the most part relatively hard and thin-bedded, corresponding to the Pläner facies of Germany. Further points of distinction are the complete absence of flints in the higher part of the northern succession and the presence within its flinty sequence of courses of continuous thick tabular flints. The faunas, too, have greater affinities with those of equivalent horizons in Germany and the USSR than with the well-

Table 6 Classification of the Upper Cretaceous in the Northern (Yorkshire and Lincolnshire) and Southern (southern England) provinces

Lithostratigraphical divisions (Northern Province)	Zones (Northern Province)	Standard zones (Southern Province)	Stages/substages		Lithostratigraphical divisions (Southern Province)
Flamborough Chalk Formation	[higher horizons present beneath Drift cover in Holderness]	*Belemnitella mucronata*	Upper Campanian	Senonian	Upper Chalk
		Gonioteuthis quadrata	Lower Campanian		
	Inoceramus lingua	*Offaster pilula*			
	Marsupites testudinarius	*Marsupites testudinarius*	Santonian		
	Uintacrinus socialis	*Uintacrinus socialis*			
Burnham Chalk Formation	*Hagenowia rostrata*	*Micraster coranguinum*	Coniacian		
	Micraster cortestudinarium	*Micraster cortestudinarium*			
	Sternotaxis [*Holaster*] *planus*	*Sternotaxis* [*Holaster*] *planus*	Turonian		Middle Chalk
Welton Chalk Formation	*Terebratulina lata*	*Terebratulina lata*			
	Inoceramus labiatus	*Inoceramus labiatus*			
	Sciponoceras gracile	*Sciponoceras gracile*	Upper Cenomanian	Cenomanian	Lower Chalk
Ferriby Chalk Formation (which includes Lower Cretaceous Red Chalk at base)	*Holaster trecencis*	*Calycoceras naviculare*			
	Holaster subglobosus	*Acanthoceras rhotomagense*	Middle Cenomanian		
		Mantelliceras mantelli	Lower Cenomanian		

known faunas of southern England on which the standard zonal scheme is based. The differences are best exhibited in the echinoids, and in the greater frequency of ammonites and inoceramid bivalves in the higher part of the northern succession. In fact, the standard zonal scheme cannot be applied in its entirety within the Northern Province, and it has proved necessary to substitute local zones in some instances. Even where standard zones have been generally adopted, the differences in zonal assemblages between the two provinces are such that correlation is comparatively crude and the majority of zonal boundaries cannot be drawn with confidence. Furthermore, since it is not possible at present to identify the horizon in the Northern Province corresponding to the base of the Upper Chalk, the tripartite subdivision in southern England into Lower, Middle and Upper Chalk is not applicable. A new lithostratigraphical classification has recently been erected for the Northern Province (Wood and Smith, 1978). This comprises, in ascending order, the Ferriby, Welton, Burnham and Flamborough Chalk Formations (see Table 6). The first of these includes the so-called Red Chalk of the previous chapter. In this Guide, however, the succession will be described within a framework of international stages and local zones. The differing classifications applied to the Chalk successions of the two provinces are shown in Table 6.

Cenomanian

Until recently the Cenomanian of the Southern Province was divided into a lower zone of *Schloenbachia varians* and a higher zone of *Holaster subglobosus*, succeeded by a zone or subzone of *Actinocamax plenus* corresponding to the Plenus or Belemnite Marls at the top of the Lower Chalk, and placed by different authorities in the Upper Cenomanian or the basal Turonian. These zones have now been replaced by a quadripartite sequence of ammonite zones which can be related to the standard substages of the Cenomanian (Rawson and others, 1978). The *Sciponoceras gracile* Zone of the new scheme, however, is not the exact equivalent of the old zone/subzone of *A. plenus* since it embraces not only the Plenus Marls but also the lower part of the Melbourn Rock up to the first occurrence of basal Turonian inoceramids of the *Inoceramus labiatus* group. The Lower–Middle Chalk boundary thus falls within the Cenomanian as defined by macrofossils.

In the Northern Province the Cenomanian Chalk is divided into three zones, in upward succession the *Holaster subglobosus*, *H. trecensis* and *S. gracile* zones (Figure 20). The boundary between the two lower zones is effectively taken at the range of thick-tested *Holaster*, and is well above the equivalent of the Totternhoe Stone, the base of which originally defined the *varians–subglobosus* boundary of the Southern Province.

The Cenomanian maintains a more or less constant thickness of the order of 21 to 24 m as far north as the Humber. Farther north condensation and thinning of individual units in the basal part of the succession can be demonstrated over the Market Weighton Block (Jeans, 1973), but in the Cleveland Basin there is a considerable increase in thickness to some 38 m at Speeton. Over most of the region the Cenomanian rests on a bored and locally stromatolite-encrusted surface of Upper Albian Red Chalk, although at Speeton

Plate 19 Chalk overlain by thin till, Selwicks Bay [TA 255 707], looking north-west from Flamborough Head. The top of the Chalk in the foreground is disturbed, probably due to cryoturbation or ice movement, and the till in the distance is landslipped. (L 1666)

Plate 20 Glacial deposits (mainly till) resting on Corallian rocks at north end [TA 128 815] of Filey Bay. The Corallian rocks comprise the Malton Oolite above the Middle Calcareous Grit. (L 1351)

there is no sharp lithological junction, and the boundary has to be drawn within a sequence of red chalks on macrofossil evidence.

With minor variations in detail the basal beds of the Cenomanian maintain their character throughout the region. The 'Paradoxica' or Sponge Bed at the base is a thin creamy yellow to pink porcellanous limestone penetrated by an anastomosing *Thalassinoides* burrow system, misidentified by early workers as a sponge and given the name *Spongia paradoxica*. This bed has not so far yielded any ammonites, the fauna being dominated by the bivalve *Aucellina coquandiana* and small terebratulid brachiopods including species of *Concinnithyris* and *Ornatothyris*. The 'Paradoxica' Bed is overlain by a bed of grey bioclastic chalk made up largely of fragmented *Inoceramus* shells. This is the lower (1st) of two distinct *Inoceramus* Beds which Jeans (1968, 1973) traced throughout the Northern Province, but which were formerly taken to be a single unit. The 1st *Inoceramus* Bed contains at its base numerous green-coated chalk pebbles, which are overlain by a concentration of small *Holaster subglobosus* with sporadic examples of a pyramidal form of *Camerogalerus cylindricus*. The pebble horizon has yielded poorly preserved *Acanthoceras* and other ammonites, indicating a Middle Cenomanian age.

The 'Paradoxica' and *Inoceramus* Beds together comprised Bed I of the succession which Bower and Farmery (1910) established for the Lower Chalk of Lincolnshire, and which is broadly applicable throughout the Northern Province. These authors drew particular attention to the occurrence of two 'bands' of conspicuously pink chalk which provide useful marker horizons in the vicinity of Louth, but tend to lose this coloration in other areas.

Above the two *Inoceramus* Beds there are two higher horizons of gritty bioclastic chalks, namely the Grey Bed (which corresponds to all or part of the Totternhoe Stone of Buckinghamshire and adjacent counties) and Bed VII of Bower and Farmery. The greyish brown colour of these chalks is readily recognisable in the field and is far more persistent laterally than the colour of the 'Pink Bands'. A generalised version of this succession supplemented by information taken from Jeans (1968, 1973) and other sources is given in Figure 20.

Between the *Inoceramus* Beds and the Grey Bed is a sequence of hard bluish grey chalk, within which there is an horizon rich in a small *Orbirhynchia* and named the Lower *Orbirhynchia* Bed by Jeans (1968). Apart from echinoids such as *Holaster subglobosus* and *Camerogalerus cylindricus*, this part of the succession yields poorly preserved ammonites including *Acanthoceras*, *Schloenbachia* and turrilitids. The Grey Bed rests with a basal concentration of glauconitised chalk pebbles on a well-defined burrowed surface, and cylindrical burrows filled with the distinctive sediment can be traced for as much as 1 m into the underlying chalk. It is very fossiliferous, characteristic species being the bivalves *'Aequipecten' arlesiensis*, *Entolium orbiculare* and *Oxytoma seminudum*, and the small belemnite *Belemnocamax boweri*. The last occurs at this horizon from Speeton to as far south as Hunstanton. The Grey Bed also yields large examples of the zonal index fossil *Holaster subglobosus*, and marks the lowest horizon from which the characteristic Northern Province/German echinoid *Echinocorys* [*Offaster*] *sphaerica* has been recorded. Overlying the Grey Bed is a thin bed of marly nodular chalk with poorly preserved Middle Cenomanian ammonites (*Acanthoceras* and

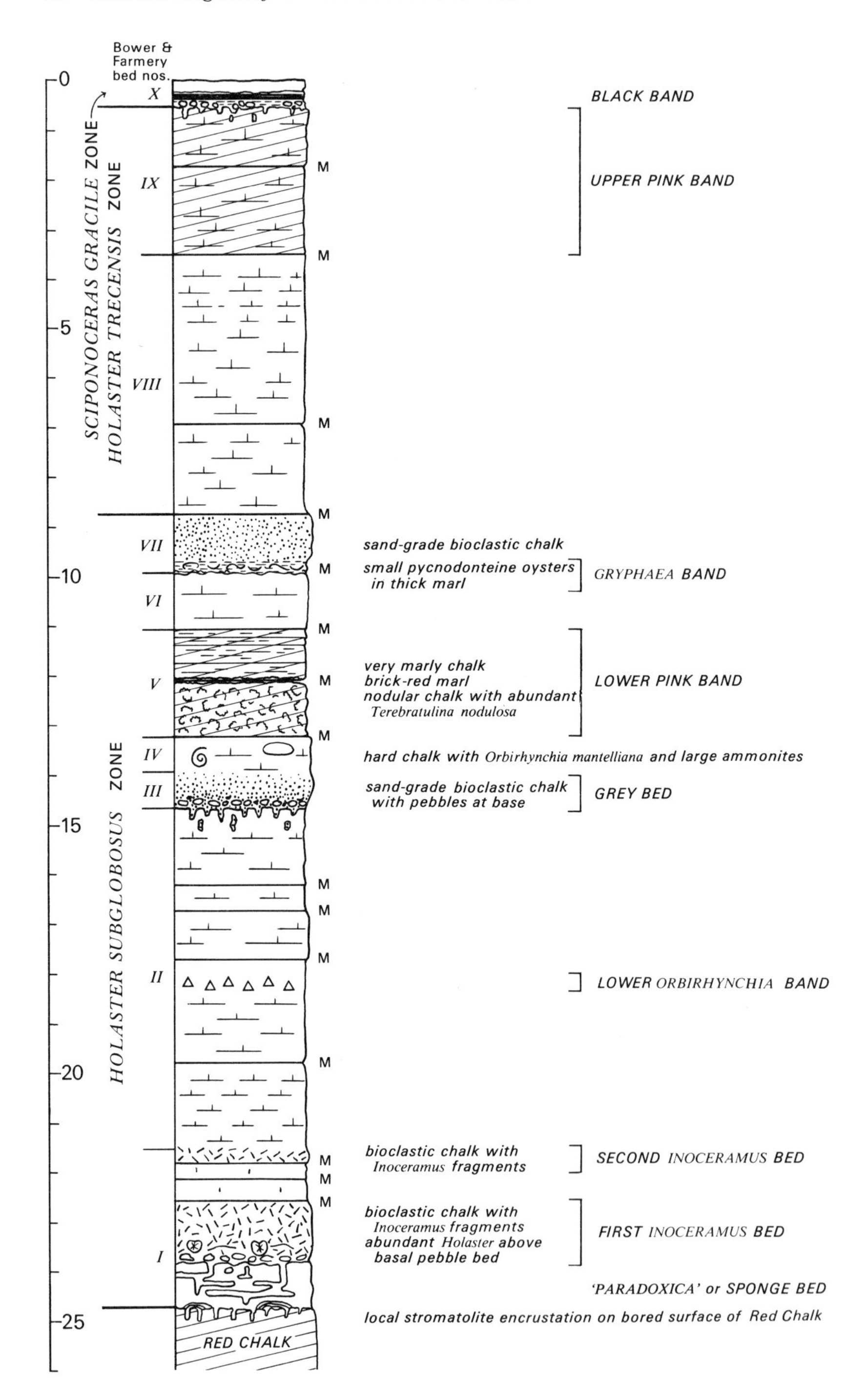

Figure 20 Section of the Cenomanian Chalk

Austiniceras) and the coarsely ribbed rhynchonellid brachiopod *Orbirhynchia mantelliana*.

The succeeding Lower Pink Band is relatively marly, and the small brachiopod *Terebratulina nodulosa* is so abundant that Bower and Farmery proposed a subzone based on it. Other fossils from this band include the serpulid *Rotularia umbonata*, the brachiopods *Grasirhynchia martini*, *Kingena concinna*, *Concinnithyris spp.* and *Ornatothyris spp.*, and echinoids including *Camerogalerus cylindricus*, *Echinocorys sphaerica* and small examples of *Holaster subglobosus*. Above the Lower Pink Band is a bed of white chalk with an indurated top.

The so-called '*Gryphaea* Band' at the base of Bed VII is a 5-cm bed of dark grey marly chalk with abundant small pycnodonteine oysters which rests abruptly on the underlying white chalk. Bed VII, which maintains a more or less constant thickness of just over 1 m, marks the upper limit of the range of *H. subglobosus*, and is accordingly taken as the top of the *subglobosus* Zone. The overlying chalk of the *trecensis* Zone, which includes the Upper Pink Band, is characterised by a thin-tested *Holaster* and *E. sphaerica* as well as depressed variants of *C. cylindricus*. Fossils at this horizon are abundant but predominantly crushed. The top of this chalk is marked by a burrowed erosion surface equivalent to the sub-Plenus Marls erosion surface of the Southern Province (see discussion *in* Jefferies, 1963).

In southern Yorkshire and northern Lincolnshire the erosion surface is overlain by a complex of variegated laminated marls collectively known as the 'Black Band' from the prevailing colour of the central part of the sequence at certain localities. At Speeton the variegated succession is thickest and includes more than one black band, closely paralleling the correlative sequences in the Münster and Lower Saxony basins of West Germany. A few kilometres to the south of Louth the Black Band as such disappears. The Black Band was formerly considered to be the correlative of the Plenus Marls of the Southern Province since *Actinocamax plenus* is recorded from several localities on both sides of the Humber, and also at Hunmanby and near Louth. However, it is probable that all these records refer to the conglomeratic bed that locally occurs immediately above the basal burrowed surface, for the Black Band itself is virtually unfossiliferous apart from the trace-fossil *Chondrites* and concentrations of fish scales. The conglomeratic bed is a rich source of fossils, notably small depressed variants of *Camerogalerus cylindricus*—formerly sold by quarrymen with residual stalks of chalk as 'fossil mushrooms'—and a diverse brachiopod fauna including *Monticlarella jefferiesi*, *Orbirhynchia wiesti* and species of *Ornatothyris*. Because this assemblage characterises high levels within the Plenus Marls, Jefferies (1963) has suggested that the Black Band is probably younger. Microfossil evidence supports this and demonstrates that the Black Band was deposited in an environment with restricted water circulation and near-stagnant bottom conditions.

The Black Band is overlain by a thin bed of buff-coloured chalk with abundant *Turnus sp.* and sporadic thin-shelled inoceramids. Above this horizon the basal Turonian bioclastic chalks with inoceramids of the *Inoceramus labiatus* group set in. This macrofossil evidence indicates that the Black Band equates with the basal part of the Melbourn Rock of the Southern Province.

Turonian

This part of the succession, comprising the equivalents of the *Inoceramus labiatus*, *Terebratulina lata* and *Sternotaxis planus* zones, is relatively poorly known, and published thicknesses are unreliable. The detailed lithostratigraphy is remarkably constant throughout the region, even down to individual flint courses. In the following account all thicknesses and stratigraphical levels relate to a standard composite succession for north Lincolnshire.

The *Inoceramus labiatus* Zone comprises hard thin-bedded and, at some horizons, pebbly bioclastic chalks with thin marly partings. Inoceramids of the *I. labiatus* group are found throughout, and flattened examples occur in profusion at certain horizons. The remaining fauna comprises small rhynchonellid and terebratulid brachiopods including *Orbirhynchia cuvieri* and *Concinnithyris sp.*; pachydiscid ammonites (*Lewesiceras*?) have been recorded from South Cave. The top of the zone is taken 5 m above the Black Band and 30 cm below the first flint course, giving a thickness of 4 m. The *labiatus* Zone is thus devoid of flint in contrast to the remainder of the Turonian succession.

No thickness can be given for the succeeding *Terebratulina lata* Zone because the top cannot be satisfactorily defined at present. Previous workers have taken the base of the overlying *planus* Zone (and thus the base of the Upper Chalk) at the level at which nodular flints were said to give way upwards to close-set continuous courses of thick tabular flints, but the palaeontological evidence for this is equivocal (Wood *in* Smart and Wood, 1976). The lowest level at which continuous tabular flints are developed is a few metres below a 15-cm marl (the North Ormsby Marl of Wood and Smith, 1978) 57 m above the Black Band. This marl is conspicuous in pits near Louth and was noted by Rowe (1929) as the 'red' or 'ferruginous marl'.

The succession conventionally assigned to the *lata* Zone comprises hard thin-bedded chalks with scattered courses of predominantly small, burrow-fill, nodular flints. These flints are relatively inconspicuous, but a notable exception is the semi-continuous course of dumb-bell shaped nodular flints about 2 m above the base of the zone. Some 1.5 m above this flint course is the 'Columnar Bed', a thin (typically 15-cm) chalk bed of variable thickness, so-called because of the very close-set vertical jointing which it exhibits. The zone is lithologically monotonous, but several thick marl seams provide useful marker horizons which can be traced throughout the region. When fresh these marls are olive-green, but they rapidly weather to shades of brown. Although their horizons can be recognised in the Yorkshire coast sections as master bedding-planes, there is no sign of the marls themselves, their absence being apparently due partly to squeezing out and partly to sea erosion.

The *lata* Zone fauna is largely restricted to *Inoceramus* of the group of *I. lamarcki* [including *I. brongniarti* J. de C. Sowerby non Mantell], *I. cuvieri*, and crushed specimens of the thin-tested echinoid *Sternotaxis planus*. Pits near Hessle were formerly well known for associated groups of crushing teeth of the ray *Ptychodus* (Plate 21), and one or more bands of a small flat-based conical variant of the echinoid *Conulus subrotundus* can be recognised here and at Burdale. The characteristic Northern Province/German echinoid *Infulaster excentricus* (Plate 21) is found in the higher part of the chalk with nodular flints, and poorly preserved examples of the diminutive zonal index brachiopod *Terebratulina lata* occur abundantly at some horizons. The *T. lata* shell is typically represented by a purple-red ferruginous oxide pseudomorph, and in some cases by an external mould.

The succession conventionally taken to be equivalent to the *planus* Zone at the top of the Turonian extends up to a thin marl (the Ulceby Marl of Wood and Smith, 1978) 70 m above the Black Band, which marks a change from thin-bedded chalks with close-set courses of tabular flints to more massive chalks with nodular flints. Some of these tabular flints contain chalk-filled cavities and locally are incompletely silicified. Fossils from this part of the sequence are predominantly brachiopods such as *Orbirhynchia* and *Kingena elegans*, but there are sporadic occurrences of the echinoid *S. planus*. In the north of the region the beds close below the marl are markedly nodular and locally intensely indurated. Where the chalk is less indurated some of the nodules are seen to be echinoids enclosed in a thin but tenaceous marl envelope. Near Louth the same beds are comparatively soft and have yielded many excellent specimens of *Micraster corbovis* (Plate 21) and a small *Infulaster*. Although echinoids are relatively common at this and other horizons in the Turonian of the region, they are very difficult to find unless they have been weathered out, and the casual collector will gain the impression that the chalk is almost unfossiliferous apart from *Inoceramus*.

The nodular chalk has yielded a number of poorly preserved ammonites of the well-known *Hyphantoceras reussianum* fauna of the hardground complex forming the Chalk Rock at or near the base of the Upper Chalk of the Chilterns. These include *Hyphantoceras* and *Lewesiceras*, and there records of other genera (eg *Allocrioceras*) from even lower horizons in the tabular flint succession. It should be noted that these ammonites are comparatively long-ranged, and that there is no evidence of Chalk Rock in the Northern Province. Wright (1935) described a molluscan fauna of Chalk Rock type with abundant small gastropods from lenses of iron-stained nodular chalk formerly exposed in the pit by the station at Kiplingcotes. Although this locality was attributed to the upper part of the *planus* Zone, it is now clear that it falls into the succeeding *cortestudinarium* Zone.

Senonian

The remainder of the Chalk in the region falls into the Senonian, a grouping which embraces the Coniacian, Santonian and Campanian stages.

Coniacian and Santonian

The boundary between the Upper Turonian *planus* Zone and the overlying Coniacian *Micraster cortestudinarium* Zone cannot be defined at present. Some 2 m above the Ulceby Marl (see above) is a 30-cm bed of marly chalk packed with small pycnodonteine oysters, with smaller numbers of brachiopods including *Cretirhynchia*, *Gibbithyris* and *Orbirhynchia*. This easily recognised marker horizon can be seen at North Sea Landing, Flamborough, and in several pits in north Lincolnshire, notably the working quarry near Ulceby. Immediately overlying the oyster bed is an horizon of locally iron-stained chalk which yields *Sternotaxis placenta* and an inflated *Micraster* near *M.*

cortestudinarium. This assemblage suggests a high *planus* or low *cortestudinarium* Zone position. Somewhat higher in the succession there is a change in flint type back to courses of lenticular and tabular flints, and much of the chalk itself is unusually soft for the Northern Province. Notable localities are the disused Enthorpe railway cutting and the famous railway station pit at Kiplingcotes, both of which yield *Inoceramus* of the *inconstans* group and echinoids indicative of the lower part of the *cortestudinarium* Zone. Higher beds in this zone are exposed in the well-known collecting localities of Willerby and Little Weighton. The fauna is rich in echinoids, of which the most characteristic are *Micraster bucailli* and a small *Echinocorys*; also present are *Infulaster excentricus*, the lowest *Hagenowia rostrata* and *Cardiotaxis aequituberculatus*. Of critical importance are large *Inoceramus* of the *lamarcki* group, and *I. schloenbachi*, the latter species being common in the upper part of the *cortestudinarium* Zone in southern England.

The junction between the flinty and flintless chalk falls within the reputedly 80 m thick *Hagenowia rostrata* Zone, which is taken to be the local equivalent of the *Micraster coranguinum* Zone, despite the fact that no definite examples of the zonal index are known. All the higher chalk seen at outcrop is completely devoid of flint.

The flinty *rostrata* Zone chalk is characterised by tabular and lenticular flints similar to those in the underlying zone, but these are replaced in the higher part of the succession by small nodular burrow-fill type flints. The topmost flints in the coast sections on either side of Selwicks Bay (Plate 19) are thin irregular tabulars which may possibly be secondary, and it is far from clear whether the flinty-flintless chalk boundary occurs at a constant horizon. Inland pits exposing the lower beds yield abundant *Inoceramus involutus*, a species which characterises the base of the *coranguinum* Zone in southern England. Notable localities are the top of the working quarry at Little Weighton and the nearby disused Eppleworth Limeworks pit where a band of these fossils coincides with a line of huge flints. A higher part of the succession is well exposed in the working quarry at Middleton-on-the-Wolds and includes a thick (5 cm) marl (the Middleton Marl of Wood and Smith, 1978). This locality is noted for lithistid sponges preserved in limonite, and *Inoceramus digitatus* J. de C. Sowerby non Schlüter occurs commonly at one horizon below the marl. Belemnites including *Actinocamax verus* and *Gonioteuthis westfalica* occur in the flinty *rostrata* chalk of several localities. *Hagenowia rostrata* also occurs but does not extend up into the flintless part of the zone.

The overlying flintless *rostrata* chalk, which is well exposed on the coast on both sides of South Landing, is characterised by an abundance of the diminutive thin-tested echinoid *H. blackmorei anterior*, formerly misidentified as the larger index species *H. rostrata*. A low level in the flintless *rostrata* Zone has yielded the basal Santonian *Inoceramus pachti*, a species which is found across Europe as far as the USSR. There is, however, no unequivocal palaeontological evidence that the boundary between the flinty and flintless *rostrata* chalk corresponds to the boundary between the Coniacian and Santonian stages, and it is quite likely that this stage boundary falls within the flinty part of the zone. Compared with southern England, the apparent absence of *Conulus* and the extreme rarity of *Micraster* in the northern equivalent of the higher part of the *coranguinum* Zone is particularly striking. Cliff sections reveal numerous small

Plate 21 Red Chalk and Chalk fossils (natural size, except 1 and 11)

1 *Inoceramus lingua* Goldfuss, $\times \frac{1}{2}$; Chalk, *Inoceramus lingua* Zone. **2** *Uintacrinus socialis* Grinnell; Chalk, *Uintacrinus socialis* Zone. **3** *Scaphites binodosus* Roemer; Chalk, *Inoceramus lingua* Zone. **4, 5** *Infulaster excentricus* (Woodward); Chalk, *Terebratulina lata* Zone, C. W. Wright Coll. **6** *Micraster corbovis* Forbes; Chalk, *Sternotaxis planus* Zone. **7** *Neohibolites minimus minimus* (Miller); Red Chalk, Middle Albian. **8** *N. m. minimus*; Red Chalk, *Mortoniceras inflatum* Zone. **9** *Marsupites testudinarius* Miller; Chalk, *Marsupites testudinarius* Zone. **10** *Ptychodus polygyrus* Agassiz; Chalk, *Terebratulina lata* Zone. **11** *Mortoniceras* (*M.*) *inflatum* (J. Sowerby), $\times \frac{1}{2}$; Red Chalk, *Mortoniceras inflatum* Zone.

fossils including corals (*Parasmilia*), but few larger fossils are seen except for belemnites and sporadic and predominantly distorted *Echinocorys*. The former include *Actinocamax verus* and *Gonioteuthis* of the *G. westfalica–G. westfalicagranulata* lineage. Giant ammonites (*Parapuzosia?*) are recorded from South Landing.

The *Uintacrinus socialis* Zone consists mostly of massive chalk and contains isolated calyx plates and brachials of the zonal crinoid (Plate 21); the cliff-section east of Dane's Dyke is one of the few English localities where complete calyces have been found. The thickness of 27 m quoted for this zone could well be an overestimate. In addition to the coast sections the zone is well exposed in the large whiting pits near Beverley and in the Bracken pit near Bainton. Both these localities include one or more bands with large specimens of *I. pinniformis*, and *Gonioteuthis granulata* is not uncommon. Thin bands packed with the oyster *Pseudoperna* [*Ostrea*] *boucheroni* are another characteristic feature of the zone, as is a form of *Echinocorys* with a truncated apex, which occurs in a band near the base of the Beverley pits.

The *Marsupites testudinarius* Zone, for which the quoted thickness of 37 m may well be too great, comprises alternations of blocky and flaggy chalk with marl bands and numerous rusty pyrite nodules. Calyx plates of the zonal crinoid (Plate 21) are common, and complete calyces can be found on the coast near Dane's Dyke and at the top of the Beverley whiting pits. Bands packed with *P. boucheroni*—including variants incorrectly recorded in earlier literature as *Pycnodonte vesicularis*—are particularly characteristic. *Zeuglopleurus rowei* occurs rarely in this and the preceding zone. The remaining fauna includes *Kingena lima*, *Orbirhynchia pisiformis*, rare *Actinocamax verus*, common *G. granulata* and fragments of large ammonites (*Parapuzosia*). *Inoceramus lingua* is markedly common at the top of the zone (Wright and Wright, 1942).

Campanian

The chalk of the *Inoceramus lingua* Zone, estimated to be about 100 m thick, is similar to that of the zone below, but the marl seams are thicker and more numerous. *I. lingua* (Plate 21) and related species occur throughout, the former being used as the zonal index instead of the echinoid *Offaster pilula*, which occurs only rarely. Bands of *P. boucheroni* are found, as in the preceding crinoid zones.

Only the lower 54 m of the zone are seen on the coast, where the famous Flamborough Sponge Beds—exposed on the foreshore between Dane's Dyke and Sewerby—yield a rich assemblage of well-preserved sponges. Characteristic species are the hexactinellids *Wollemannia laevis*, *Rhizopoterion cribrosum*, *Sporadoscinia strips* and *Leiostracosia punctata*; the dominant lithistids include species of *Amphithelion* [*Verruculina*], *Aulosoma*, *Laosciadia* and *Siphonia*. This horizon also yields a large dome-shaped *Echinocorys*, sporadic examples of *Offaster pilula*, an undescribed *Cardiotaxis* and the asteroid *Metopaster stainforthi*. An *Orbirhynchia* occurs abundantly and there are sporadic examples of the bivalve *Oxytoma tenuicostatum*, which is a widely distributed Lower Campanian fossil in northern Europe, particularly in the USSR. *Gonioteuthis granulataquadrata*—including very large specimens—and *Actinocamax verus* occur throughout these beds, and the ammonite *Hauericeras pseudogardeni* is not uncommon. A single specimen of the rare *Belemnello-camax grossouvrei* is known from Ruston Parva.

Higher beds, said to be 46 m thick, of the *lingua* Zone are exposed in pits on the Wolds near Bridlington and around Driffield. They yield heteromorph ammonites, including *Scaphites binodosus* (Plate 21), *S. inflatus*, a *Glyptoxo-ceras* and indeterminate baculitids, and have been separated as a *Scaphites binodosus* Subzone. Belemnites recorded at White Hill, near Bridlington, include *Gonioteuthis quadrata, A. verus* and *Belemnitella sp.* This subzone is characterised by the abundance of the cirripede *Zeugmatolepas cretae* and a small smooth thin-shelled pectinid recorded as *Entolium* [*Syncyclonema*] *orbiculare*. An undescribed *Pseudoptera* also occurs, and *Oxytoma tenuicostatum* is much commoner at this horizon than in the lower part of the *lingua* Zone. Pits at Bessingby have yielded several examples of a small *Micraster* with a deep and very narrow anterior sulcus.

It is usually considered that the *lingua* Zone corresponds to the whole of the *pilula* Zone of southern England, but that those *binodosus* Subzone localities yielding *Gonioteuthis quadrata* might belong to the lower part of the overlying *quadrata* Zone. *G. quadrata*, however, is not restricted to the zone of that name, but also occurs in the German equivalent of the higher part of the English *pilula* Zone. The occurrence of both *Scaphites binodosus* and inoceramids of the *I. lingua* group clearly point to a *pilula* Zone position for the highest beds of the exposed Yorkshire succession.

Higher parts of the Chalk succession are present beneath the Drift cover of Holderness, where the total thickness of Chalk reaches 500 m. Microfaunal evidence from boreholes here indicates that the top of the succession falls into the Upper Campanian (*Belemnitella mucronata* Zone). Good specimens of the zonal index *B. mucronata* are said to be frequently found in the Drift of this area. Although total Chalk thicknesses of up to 800 m are known offshore on the landward side of the Dowsing Fault, there is no evidence of Maastrichtian Chalk.

10. Tertiary

The only rock of undoubted Tertiary age within the region is an igneous intrusion—the Cleveland Dyke. Tertiary sediments, however, floor much of the North Sea, and the edge of these, unconformably overlying Mesozoic rocks, reaches to within 70 km of the Yorkshire coast (see Plate 1).

Tertiary sediments

The end of the Cretaceous Period was marked by a major regression of the sea. In the earliest Tertiary stage—the Danian—the flooded area and consequent deposition did not extend farther west than the central part of the present North Sea, so that wide areas of pre-Danian Chalk, including that of eastern England, were exposed to erosion. In the post-Danian Palaeocene a flood of detrital sediments was introduced into the basin from the north, and the area of marine deposition again expanded, reaching a maximum during the Eocene. The Oligocene rocks were less widely deposited than Miocene and Pliocene rocks, which in places rest unconformably on the older Tertiary sediments.

How close to the present coast Tertiary strata accumulated is a speculative matter, but in general in the North Sea it is thought that they have not been eroded back very far, since piecemeal attenuation and lithological characters indicate proximity of contemporary shorelines to present outcrops. The Tertiary sediments, thought to be Palaeogene (probably Eocene), which occur north-east of Whitby are mainly mudstones resting with marked unconformity on gently folded Upper Cretaceous rocks (Dingle, 1971).

By analogy with other parts of the North Sea Basin and in view of thickness trends, it seems unlikely that any significant thickness of marine Tertiary beds was deposited within the present land area, although survival of relics (perhaps of Neogene age) remains a possibility. In general Yorkshire and Lincolnshire were subject to steady removal of Mesozoic strata as progressive uplift and easterly tilting brought successively older beds within range of subaerial erosion during the Tertiary Period. Erosion surfaces, some of which are presumed to be of Tertiary age, are dealt with in Chapter 12.

Cleveland Dyke

The Cleveland Dyke of Yorkshire is the south-eastern portion of one of the dykes related to the Tertiary igneous complex of Mull. It is almost certainly continuous at depth with the Armathwaite Dyke of Cumbria, which can be traced north-westwards on the aeromagnetic map across the Solway Firth.

Within the region the dyke cuts Triassic and Jurassic rocks and is up to 25 m wide. It is first seen in the north near Eaglescliffe Station in the Tees Valley and can be followed south-eastwards in a series of *en echelon* outcrops for about 50 km to Fylingdales Moor.

The dyke rock is a tholeiite, consisting of phenocrysts of plagioclase and pigeonite set in a glassy or felsitic groundmass which contains augite and quartz. Adjoining sediments are affected only within 2 m of the dyke, the alteration consisting of discoloration, spotting and induration. Potassium-argon isotopic age determinations give a minimum age of 58.4 ± 1.1 Ma (Eocene).

The Cleveland Dyke has been extensively worked, almost to exhaustion, in quarries and shallow mines, chiefly for use as roadstone. The Nature Conservancy Council are preserving parts of a quarry in this rock near Great Ayton as a site of special scientific interest.

11. Structure

Major structural features

Throughout the Carboniferous, the region lay within a broad zone of depositional basins which extended from Ireland eastwards across northern Europe. This zone was separated from the depositional area which later became the severely folded belt of the Hercynian orogeny by the Wales–Brabant massif, but within it there was extensive intra-Carboniferous differential vertical movement and some late Carboniferous folding. These movements led to the rise of the Pennines, which appear to have remained a positive feature, subject to minor uplift, from the Permian to the present time.

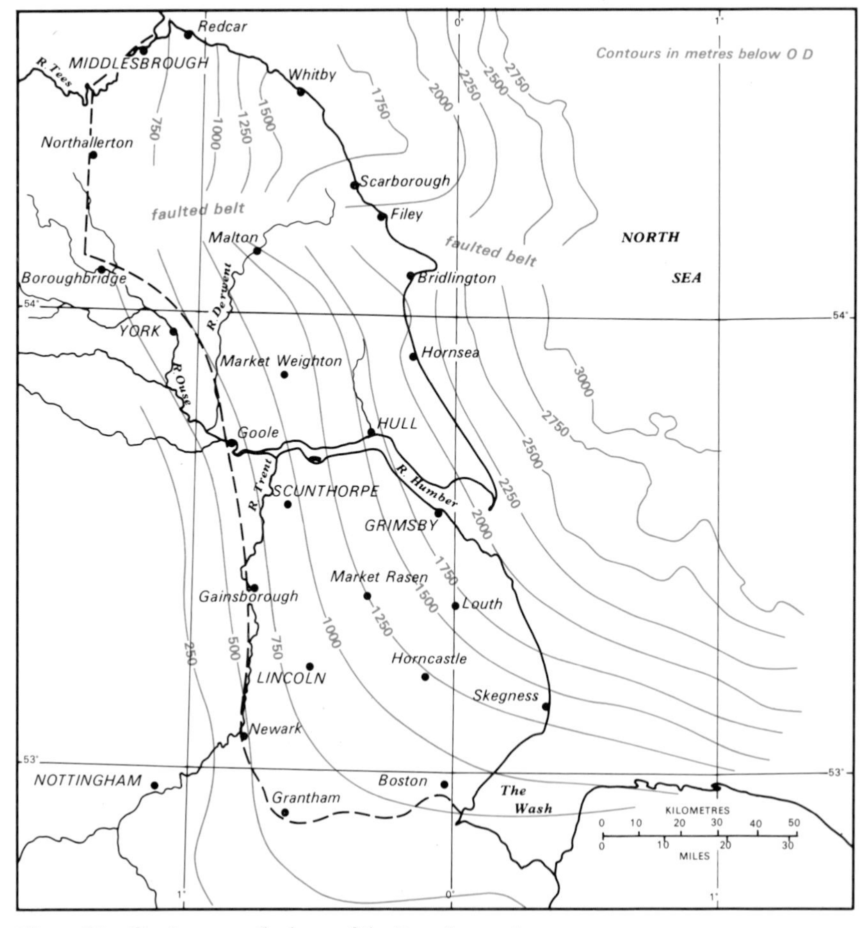

Figure 21 Contours on the base of the Permian rocks

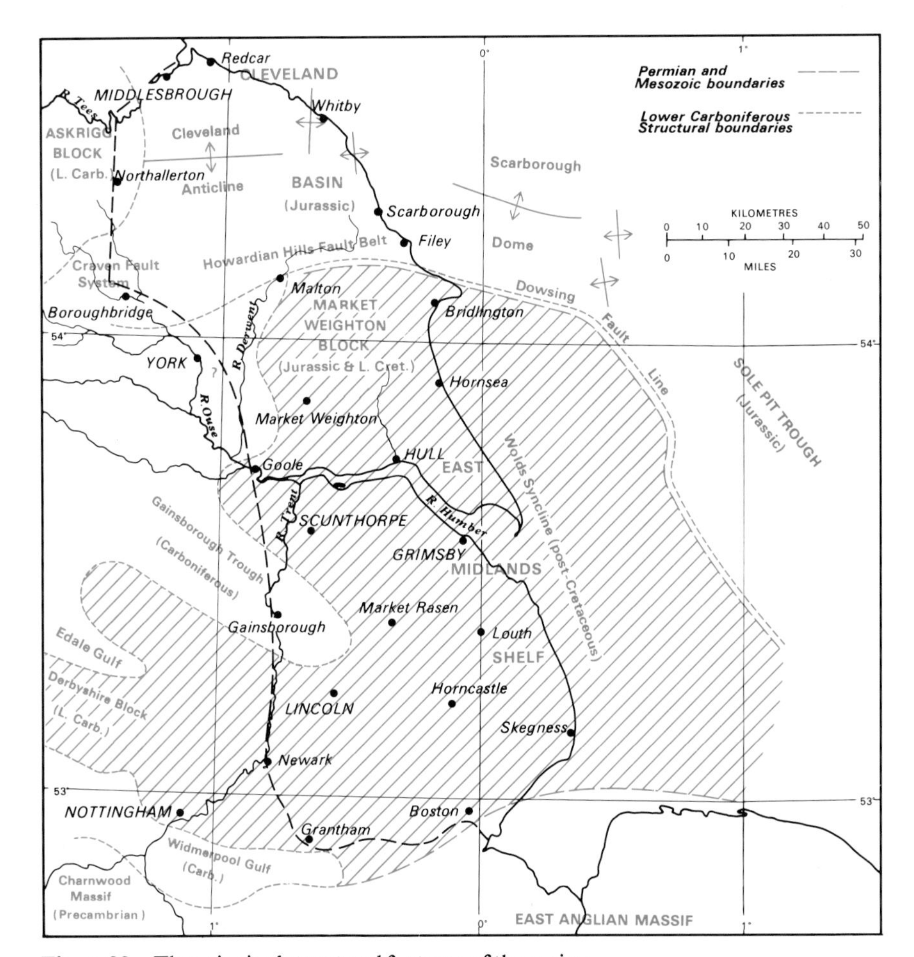

Figure 22 The principal structural features of the region

From the Permian onwards the dominant feature was epeirogenic subsidence of the North Sea Basin, which came into existence in Rotliegendes (Lower Permian) times as a north–south feature between the Palaeozoic highs of Scotland and the Pennines in the west and Norway and the Hercynian massifs in the east. Throughout the Permian and Mesozoic and on to the present day the Tees–Wash region thus formed a marginal part of the Southern North Sea Basin (Figure 21).

The region is nevertheless divided into two parts which have dissimilar structural histories—the boundary varying in time between the west–east Howardian Hills fault belt and the Market Weighton area (Figure 22). To the north and continuing out to sea is the Cleveland depositional basin, which was a major downwarp in the Lower Carboniferous and Jurassic, but which was uplifted in end-Carboniferous and end-Cretaceous times. To the south is the East Midlands Shelf, its northernmost extremity formed by the notably rigid Market Weighton Block. This southern area was relatively buoyant in the Lower Carboniferous and Jurassic, but escaped the end-Carboniferous uplift

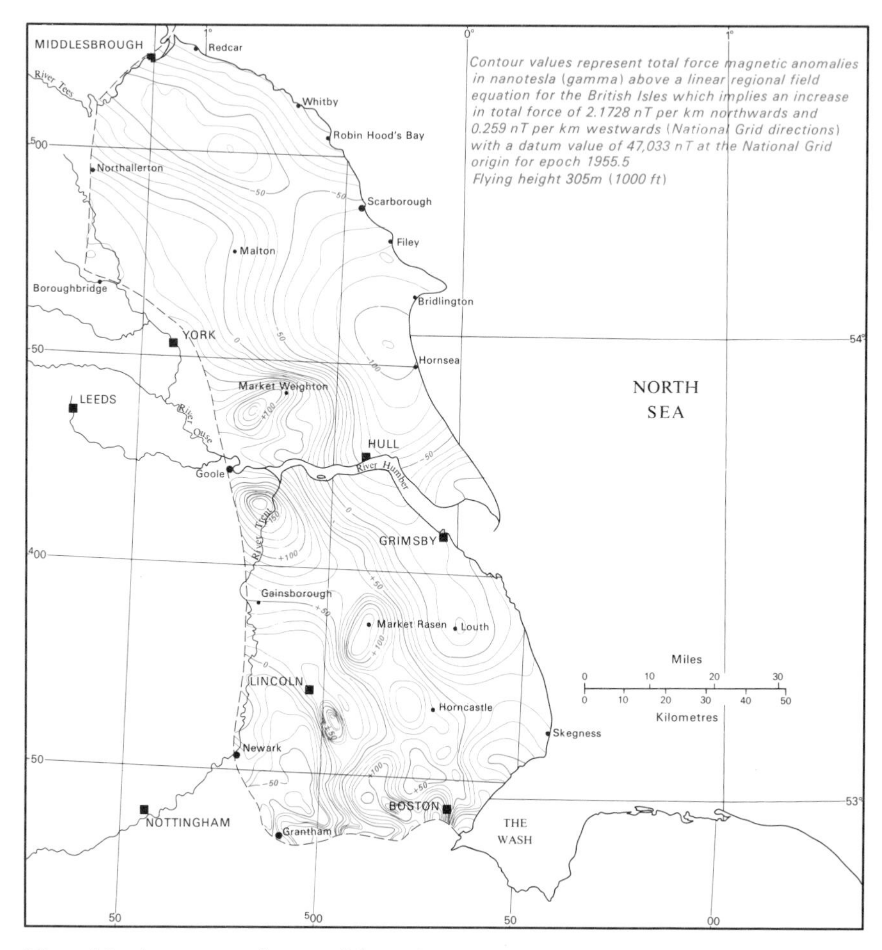

Figure 23 Aeromagnetic map of the region

and also the end-Cretaceous inversion which affected the Cleveland Basin and the offshore Sole Pit Trough. The contrast between these two provinces is illustrated by the map of magnetic anomalies (Figure 23), which shows a more simple pattern of contours in the north than from Market Weighton southwards, reflecting deeper magnetic basement in the basinal area. The boundary between these provinces, which continues south-eastwards in the Dowsing Fault Belt, has been a major lineament in British structure.

Within the Cleveland depositional basin, the North Yorkshire Moors coincide with a general east–west trending broad arch which carries subsidiary domes (Chop Gate, Danby Head, Eskdale and Robin Hood's Bay—Figure 24). Offshore the Cleveland Basin extends some 65 km eastwards, with the arch continued eastwards by the very similar broad Scarborough Dome (Figure 22) which produces a Triassic inlier 30 km north-east of Scarborough (Plate 1), and by other, more irregular, domes within the Chalk outcrop still farther east—the Blake, Fox-Strangways, Lamplugh and Jukes-Browne

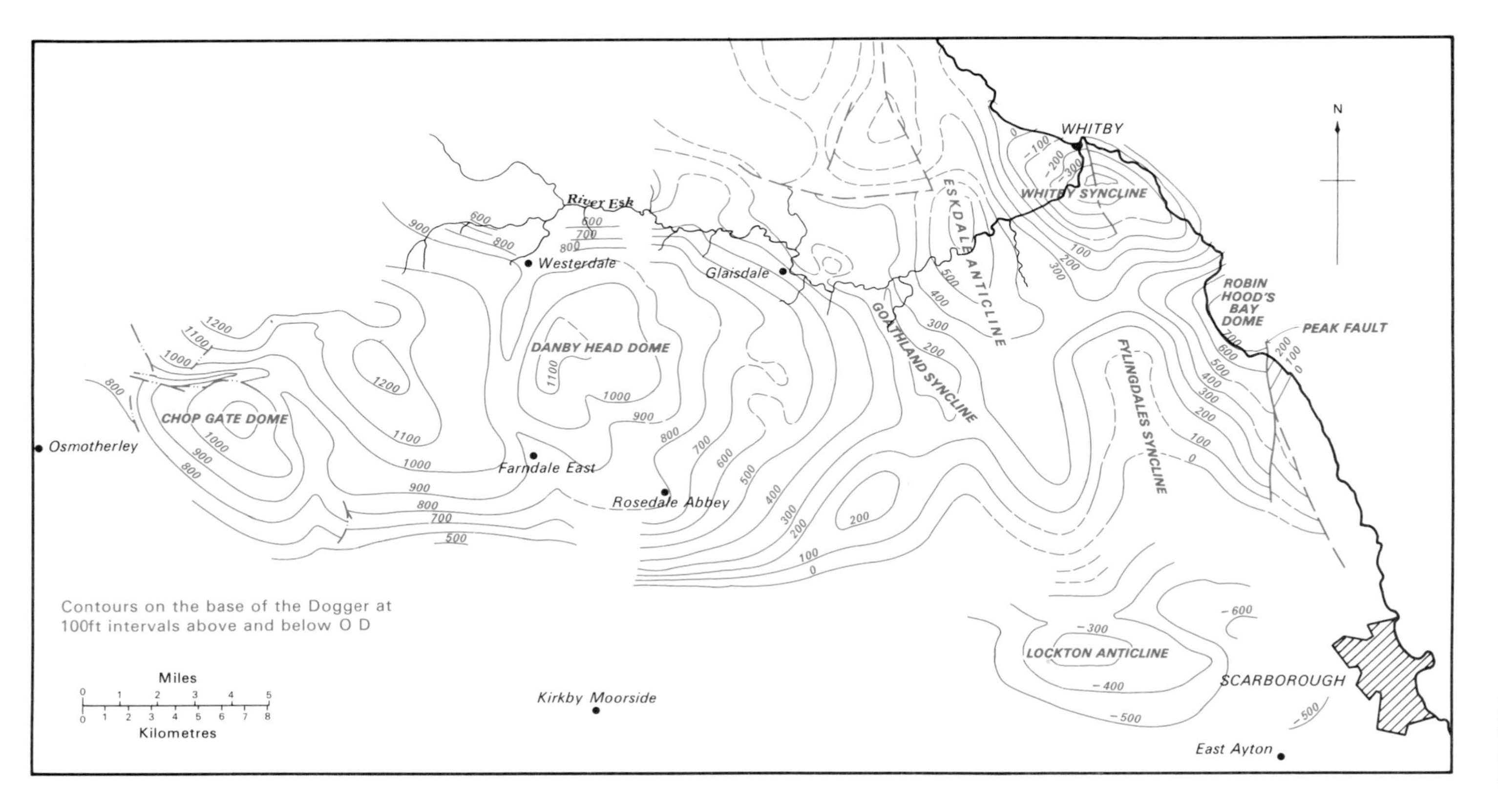

Figure 24 Subsidiary folding of the Jurassic rocks of the North Yorkshire Moors

Domes which Dingle (1971) defined and named after notable Yorkshire geologists of the last century.

In the North Sea the faulted edge of the East Midlands Shelf is coincident with the edge of the main Permian salt basin, with its thick salt subject to large-scale flowage and the inception of salt pillows and piercement plugs, some of the displaced salt overlapping on to the shelf edge (Brunstrom and Walmsley, 1969).

The edge of this basin, The Sole Pit Trough, was the locus of particularly heavy sedimentation in the Lower Jurassic and was a deeply subsided depositional trough which subsequently suffered inversion, the original downwarp being followed by an anticlinal uplift. There is evidence that this inversion also affected the Cleveland Basin, and was largely responsible for the rise of the broad Cleveland Anticline and of its eastward offshore homologues, mainly in pre-Tertiary times. The evidence for this is described below.

The Market Weighton Block

The original mapping of Yorkshire demonstrated the very pronounced over-step of the Upper Cretaceous rocks across successively lower horizons towards the Market Weighton area from both north and south, effectively separating the Jurassic outcrops of north Yorkshire from those of the Midlands (Plate 1). Kendall in 1905 was the first to emphasise that this separation had been produced by intermittent uplift of the intermediate area, and from the relationship of an 'upfold' separating two basinal areas described the structure as an 'anticline', supporting this concept by widely quoted (although misleading) diagrammatic sections. His view of the recurrent relative uplift has been fully endorsed by later work, but the anticlinal interpretation has been modified; the area with attenuated deposits is essentially an unfolded block, 20 km across from north to south (Kent, 1955; Bisat and others, 1962). This remained buoyant—with only minor periods of erosion—through much of Jurassic time while both the Cleveland area and the more southerly parts of the East Midlands Shelf were undergoing steady subsidence (Figure 25). Positive uplift occurred intermittently (with consequent erosion of accumulated sediments), but this was relatively insignificant compared with the regional subsidence.

Kendall (1905) regarded the trend of the structure as east–west and continuous with the Wharfe anticline, but the latter concept is rendered unlikely now that we know that the area retains its cover of Coal Measures, which are stripped off farther north. Data from deep boreholes which are now available show, however, that the east–west trend is indeed broadly dominant, the attenuated belt continuing eastwards to a line beyond the present Holderness coast (see Figures 9, 10, 14, 15 and 19).

There is a decided north–south asymmetry across the block produced by the sharply faulted downwarp of the Cleveland Basin to the north as compared with the relatively gentle southward thickening into Lincolnshire on the East Midlands Shelf. This has lately led to the concept that the Market Weighton Block is essentially a hinge between the shelf and the northern basin, but it is strictly the abutment of a complex hinge, with the main line of inflexion on the north changing through the Jurassic (tending to shift northwards), and it is associated also with complex faulting.

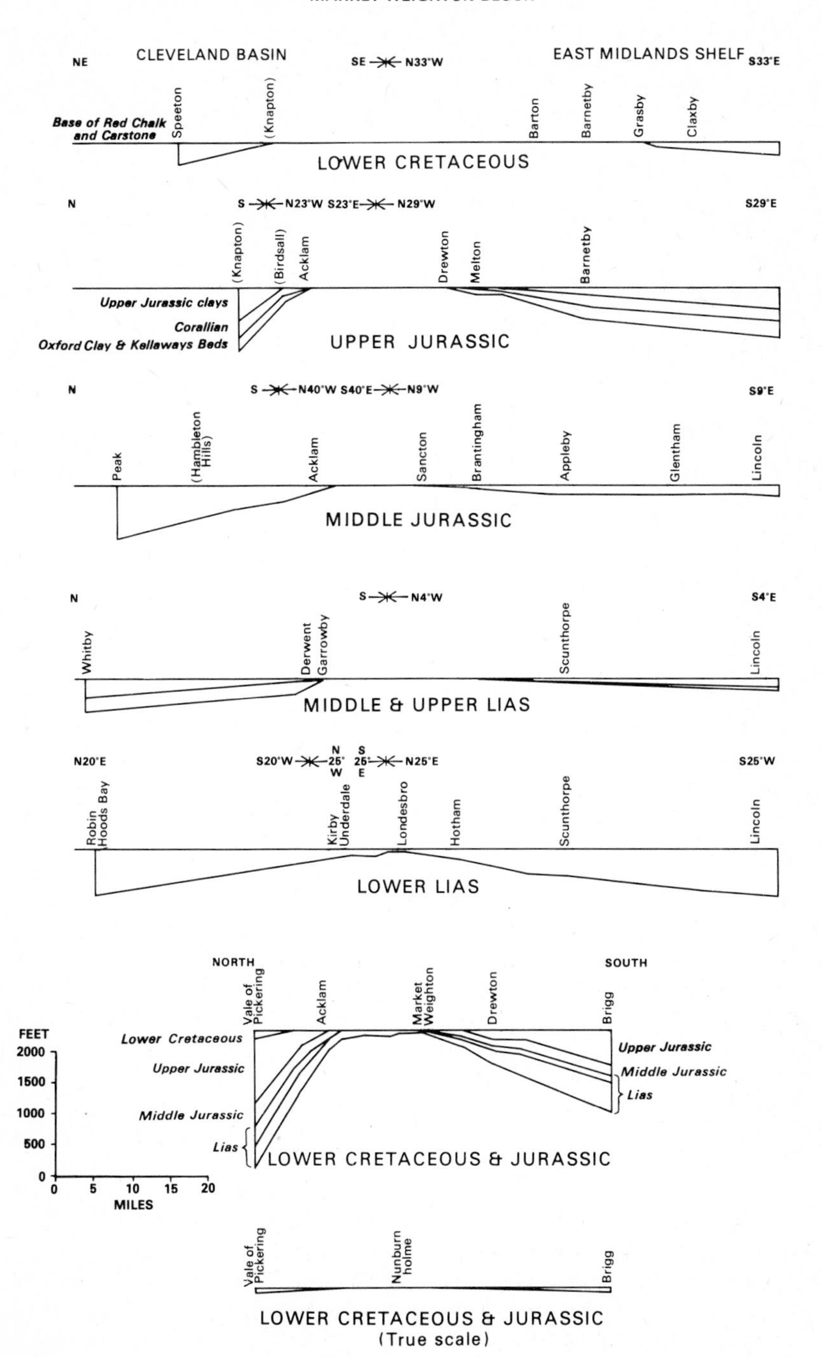

Figure 25 Diagrammatic sections across the Market Weighton Block during the Jurassic and Lower Cretaceous

There are enough boreholes through the Permian to establish that salt (halokinetic) movement was not a factor in producing the Market Weighton Block. It is seen regionally as the shelf edge, possibly given buoyancy and rendered particularly resistant to flexing by granite intrusion at depth. There is now gravity evidence of the existence of such a granite.

There is as yet little evidence of relative movement of the Market Weighton Block in Permo-Triassic times although there are unexplained anomalies in the Permian rocks associated with the Howardian Hills faults. The Mercia Mudstone (Keuper Marl) is the oldest rock to show some thinning, for the Sherwood Sandstone (Bunter) actually thickens across the area. Rocks of the Penarth Group (Rhaetic), although thinner than in Lincolnshire, are continuous across it. The lower part of the Lower Lias (*planorbis* to *angulata* zones) is well developed, although locally there is a recurrent tendency to unusually shallow-water conditions (limonite ooliths and occasional pebbles). The upper part of the Lower Lias and the Middle Lias thin steadily northwards in Lincolnshire and are relatively poorly developed in the Market Weighton area, though their attenuation is in part due to late or post-Toarcian uplift and erosion. The Dogger fails over the Market Weighton Block, but the later marine units equivalent to the Lincolnshire Limestone were originally persistent, at least in the west, although they are now limited to pre-Cretaceous synclines. The Great Oolite thins out northwards in Lincolnshire near Brigg and the Cornbrash dies out on the Humber. The Kellaways Sand is well developed near South Cave, but the Oxford Clay is thin and locally cut out, and its removal in the Market Weighton area probably dates from an early Oxfordian phase of uplift and erosion. The Langdale Beds (Middle Callovian) of the Scarborough area contain a derived Bathonian (and possibly earlier) microflora, which seems most likely to have originated from erosion of the edge of the Market Weighton Block to the south (M. D. Muir and W. A. S. Sargeant, in litt.). The Ampthill Clay is of appreciable thickness on the Humber and in the Acklam area (Brown Moor Borehole), indicating that deposition was largely unaffected by any movement of the Market Weighton Block and was probably continuous across it. Later Jurassic beds were removed by the uplift which preceded deposition of the Albian Carstone and Red Chalk.

Most of the uplift can be accounted for by the amount of thinning of the sedimentary units, rather than by the thickness of material removed by erosion.

The Cleveland – Sole Pit Trough Basin

The existence of areas that were initially basins but subsequently underwent a rebound—'inversion structures'—has been widely recognised in Europe (Ziegler, 1975). In England the Wealden Anticline is a well documented example: it was the site of a major Jurassic downwarp which is now anticlinal at surface as a result of inversion, but since the amount of the later central uplift was less than the earlier basinal subsidence, the base of the Jurassic remains gently synclinal. The Cleveland Basin and its offshore continuation as the Sole Pit Trough is a comparable case, but one in which part of the Palaeozoic history is also known.

Inversion of the Sole Pit area was deduced by Heybroek (1975) by analogy with the Dutch Graben, and Hancock and Scholle (1975) provided supporting evidence based on the fact that the Upper Cretaceous is thin over the trough but abnormally thick on the eastern flank. This is in contrast to the Lias, which measures over 1000 m—three times the normal thickness—in the trough, as it does in the Jurassic troughs in Germany which were also subsequently inverted (Brunstrom and Walmsley, 1969). There is, additionally, palaeothermal evidence indicating that the Sherwood Sandstone beneath the Sole Pit Trough was formerly buried much more deeply than at present (Marie, 1975).

Realisation that the Cleveland Anticline is analogous to the Weald was initially derived from analysis of aeromagnetic data (Kent, 1974). The northern edge of the original basin is now missing, eroded away on land, but the regional northward thinning is established north-east of the Tees in the Tyne Hole area of the North Sea (Dingle, 1971). It may be deduced that the anticlinal uplift was largely intra-Jurassic and intra-Cretaceous. A mainly mid-Tertiary age for the folding of Cleveland was formerly assumed from physiographical evidence, but the displacement of Tertiary surfaces is on a much smaller scale than the amplitude of the main Cleveland fold. The relationship of the basal Tertiary unconformity to the Mesozoic folds as mapped by Dingle (1971) offshore (Plate 1) provides evidence of an earlier date for the Cleveland folding in the offshore area.

The mechanism controlling inversion structures is not understood. Salt movement was clearly not the critical factor, for the Permian salt is relatively thin in Yorkshire and there is no salt in the Weald. More northerly North Sea Mesozoic basins have not undergone inversion. The alternatives are orogenic stress (a precursor of the Alpine movements) which compressed the more southerly basins until they 'locked' tight and then ceased to respond, or a special kind of isostatic rebound. The long period over which the mechanism operated tends to favour the latter explanation, but the coincidence with the late Carboniferous reversal of displacement of the trough-shelf boundary also has to be taken into account.

Other structural features

In addition to the broad structural units described above there are a number of smaller or subsidiary structures of local importance.

Near the Yorkshire coast north–south trends are notable and minor structures, eg the Eskdale Anticline, produce segmentation of the main Cleveland Anticline (Lees and Cox, 1937). The Robin Hood's Bay Dome is more equidimensional, but it is closed on the seaward side by the north–south Peak Fault, across which there is a sudden increase in thickness of the Upper Lias due to the incoming of Yeovilian strata, which are absent on the west. Facies differences at other horizons are recognised on opposite sides of the fault. This fault was originally ascribed to contemporaneous intra-Jurassic vertical movement, but it is now generally considered to be more probably a transcurrent displacement, juxtaposing a thicker sequence from a different (perhaps more northerly) part of the offshore basin against that of the present inland development (Hemingway, 1974).

From the Vale of Pickering to the Humber the dominant trends are east–west, with the Howardian Hills fault belt as the main feature. The Helmsley–

Filey Fault on the northern side of the vale is probably the most northerly of this series, although it is not suspected (as are some of those farther south) of intra-Jurassic movement. There are small faults crossing the Market Weighton area. Near Garrowby and Givendale these are east–west, one at Millington

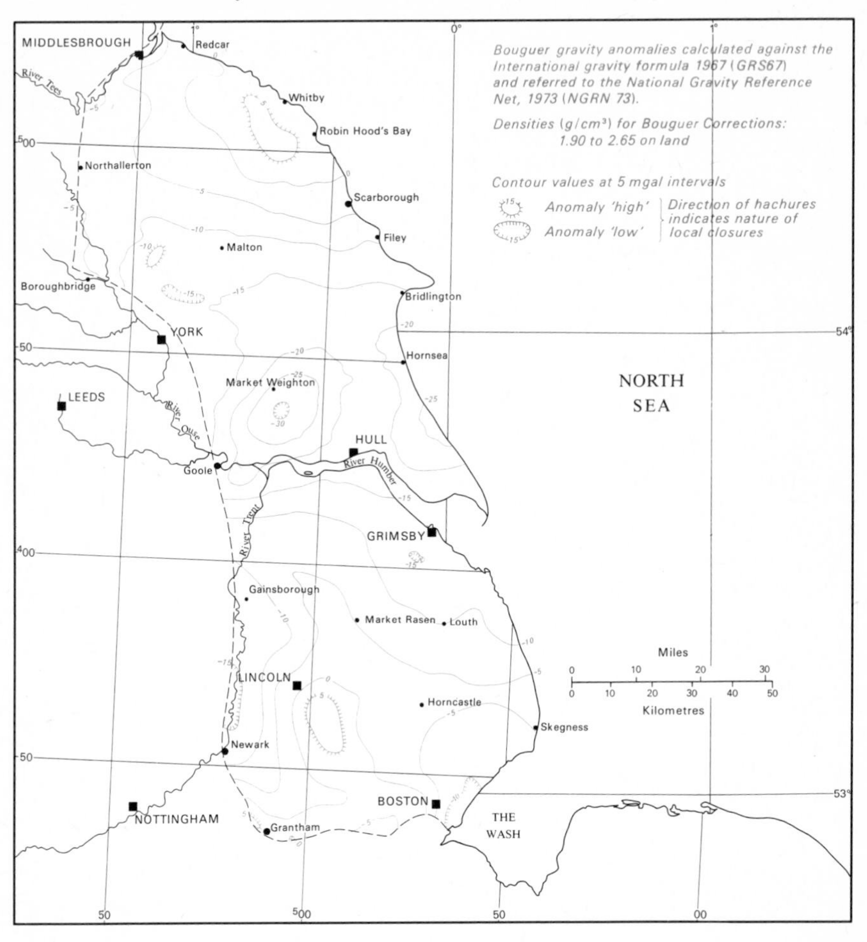

Figure 26 Bouguer gravity anomaly map of the region

trends north-east–south-west and there are others not yet adequately mapped. These fractures are in part associated with shallow east–west folds in the Cretaceous mapped by Versey (1923).

In Lincolnshire, the west-north-westerly Flixborough–Brigg–Caistor structure produces deflections of both Jurassic and Cretaceous outcrops. It is associated with faults, which have an important effect on the mining of the Frodingham Ironstone. Sixteen kilometres farther south a minor west-north-westerly anticline near Atterby affects the Lincolnshire Limestone outcrop, and the absence of the Upper Lincolnshire Limestone in its crestal area at Snitterby may be due to contemporaneous uplift. Additionally, between Hemswell and Glentworth a double bend in the Lincolnshire Limestone scarp marks the line of a larger west-north-westerly fold, which may have undergone contemporaneous movement because in Spital in the Street No. 1 Borehole the Upper Estuarine Beds rest directly on the Lower Lincolnshire Limestone (Kirton Shale), the Hibaldstow Beds being again cut out. This fold is coincident with the easterly continuation of the deep-seated Askern Anticline, which flanked the Lower Carboniferous Gainsborough Trough and was the site of a major post-Coal Measures uplift. It presumably reflects a small renewal of differential movement in Middle Jurassic times. The two more northerly structures (Snitterby and Flixborough) may be similarly related to Carboniferous structures. A widening of the Lincolnshire Limestone outcrop in the Nocton area south of Lincoln is coincidental with a deep-seated pre-Permian high (the Nocton Block) which seems to have been intermittently buoyant during the Jurassic (eg during the deposition of the Middle Lias and Lower Lincolnshire Limestone). Over this uplift the Carboniferous rocks are much thinner than in the basins to east and west, and marked magnetic and gravity anomalies (Figures 23 and 26) indicate that the dense basement rocks are here relatively shallow.

In south Lincolnshire gentle east–west folding is apparent. Just south of Sleaford minor faults mark the trend of an east–west monocline across which the gentle eastward dip of the Jurassic rocks changes abruptly (Swinnerton and Kent, 1976) from 1–2° in the north to less than 1° in the south, a factor of considerable importance for the availability of underground water in the aquifers beneath the Fens. Because of the gentle dips and shallow east–west folding to the south of this line, the outcrop pattern changes from the marked linearity of north and mid-Lincolnshire to the relatively complex patterns of the Midlands.

12. Quaternary

During the Tertiary Period the region had been uplifted, producing an easterly tilt towards the North Sea Basin, which was correspondingly depressed. In the absence of contemporaneous deposits on land, attempts have been made to deduce the Tertiary and early Quaternary denudational history and drainage evolution of the region from physiographical evidence. A summit surface, regarded as the relic of an old peneplain, and several lower planation surfaces have been recognised, some of which have been attributed to marine, and others to subaerial, agencies. Ages ranging from early Tertiary to early Quaternary are claimed for them, and the configurations of some of the older, higher, surfaces suggest subsequent tilting and flexuring. The elevations of the named surfaces are given in Table 7.

The Quaternary Period, variously estimated to include the last 1.5 to 3.5 million years, witnessed marked climatic changes, the colder phases producing glacial and periglacial conditions. The resulting eustatic and isostatic effects on sea level gave rise to a complex history of erosion and deposition in the valleys and along the coast. The region was invaded at various times by ice sheets. On the higher ground of the region the most obvious evidence of glaciation is physiographical. Here pre-existing drainage was modified by powerful flows of meltwater which cut substantial channels beneath the ice and beyond its margins. Lower ground was covered by extensive glacial and associated deposits, which buried the pre-existing topography.

Tills are the most widespread and characteristic of the glacial deposits, being heterogeneous accumulations of erratic rock fragments commonly transported from a considerable distance and including large boulders, set in a clayey, silty or, more rarely, sandy matrix. Lodgement till is deposited at the base of an ice sheet, melt-out till forms within, and ablation till on the surface of, stagnating ice, and flow till results from solifluction of supra-glacial debris down an ice front. Sand and gravel, also common in the region, are

Table 7 Planation-surface elevations in the Wolds (with generalised heights in metres above OD)

Yorkshire *After Lewin, 1969*	Lincolnshire *After Straw, 1961a*
Wintringham–Huggate (about 200)	
Ganton–High Hunsley (about 165) Langtoft–Hessleskew (120 to 145)	High Street–Bluestone (135 to 165)
	Kelstern (116 to 128)
Kilham–Rowley (90 to 114)	Beelsby Top–Haugham (85 to 105)
Grindale–Kiplingcotes (52 to 75)	Burnham (58 to 73)
Lockington (about 45) Etton–Garton (30 to 45)	Brocklesby Park (24 to 45)

deposited by meltwaters both within and beyond the ice margins, and around the edges of extra-glacial lakes in which thick deposits of laminated clay and silt accumulated. Periglacial features, including cryoturbation structures and patterned ground resulting from repeated freezing and thawing of ground ice, solifluction (Head) deposits, some of the blown sand, rare loess and scattered ventifacts, were formed during the colder phases in areas not covered by ice. It is probable that much of the cambering and valley-bulging in the deep valleys was facilitated by periglacial conditions. Sediments formed in the warmer phases are relatively rare, but within the region there are several small deposits containing plant debris, mollusca, insects or mammalian remains indicative of a temperate climate. Since the end of the last cold phase, thick fresh-water, estuarine and marine deposits, ranging from clays to gravels, have accumulated in the valleys and along the coast.

Subdivision of Quaternary time is based on recognition of alternating cold (glacial) and warm (interglacial) stages. That part of the British classification which is pertinent to the region is given in Table 8. Absolute dating beyond the limits of radiocarbon assay (in practice about 50 000 years) is extremely difficult, and estimates of the age of the Anglian Stage vary from about 200 000 years to more than half a million years.

Table 8 Quaternary stages recognised in the region*

British stages	NW European stages	Climate	Traditional northern English terms
Flandrian	Flandrian	temperate	'post-glacial'
Devensian	Weichselian	cold (glacial)	'Newer Drift'
Ipswichian	Eemian	temperate	(not differentiated)
Wolstonian	Saalian	cold (glacial)	'Older Drift'
Hoxnian	Holsteinian	temperate	(not differentiated)
Anglian	Elsterian	cold (glacial)	'Older Drift'

Several earlier stages are represented by deposits in East Anglia and elsewhere, but no pre-Anglian deposits have been recognised in the region

*There is some doubt about the validity of the Wolstonian Stage (particularly as a glacial episode) and therefore about pre-Ipswichian correlations with NW Europe. In the text however, pre-Ipswichian deposits are given the ages previously attributed to them.

Pre-Ipswichian glacial stages

The wide distribution of the older glacial deposits indicates that ice covered the entire region on at least one occasion. Although these deposits are demonstrably pre-Ipswichian, there is no evidence, except possibly at Speeton, Kirmington and Welton-le-Wold, to determine whether they are of Anglian or Wolstonian age, or both.

In north-east Yorkshire these older glacial deposits are limited to a few small outcrops of till, sand and gravel occurring mainly around the Vale of Pickering (Figure 27), together with relict patches and scattered erratics on the North Yorkshire Moors and Yorkshire Wolds. Most of the erratics are of indeterminate derivation, being composed of quartz, quartzite or chert, but a few are recognisably from the Lake District, Pennines or north-eastern England. The best exposures of older glacial deposits are along the Holderness coast, where the Basement Till crops out beneath Devensian tills at Dimlington Cliff and between Bridlington and Sewerby, with outliers on the Flamborough peninsula (Catt and Penny, 1966). The Basement Till is a grey clay, in places with a green tinge, containing erratics mainly derived from north-eastern England but including a few examples of larvikite and rhomb porphyry from Norway. These latter erratics were probably picked up in the North Sea from older glacial deposits or contemporaneous Scandinavian ice. Orientation of stones, mineral grains and small folds in the till suggest that the ice approached Holderness from the north-east (Penny and Catt, 1967). At Bridlington and Dimlington the Basement Till contains irregular inclusions of glauconitic sand ('Bridlington Crag') and bluish grey clay ('Sub-Basement Clay') respectively; both contain marine molluscs, indicating that they are masses of sea-floor sediment picked up by ice. Since the Basement Till is generally considered to be of Wolstonian age, these masses have been presumed to be Hoxnian, but they have yielded dinoflagellate cysts which suggest one of the pre-Anglian stages not otherwise represented in the region.

Lincolnshire possesses extensive spreads of pre-Ipswichian tills (Straw, 1969), with minor associated sand and gravel deposits (Figure 27). The tills crop out mainly on elevated ground, but are rare on the highest ground of the Wolds and Lincoln Edge. Both their clayey matrices and erratics indicate local derivation from the north. The Calcethorpe Till occurs sparsely on the Wolds but extensively around the Bain Valley, and is predominantly chalky with abundant flint. To the west the Belmont Till is largely derived from Lower Cretaceous rocks. Farther west, in the Ancholme and Witham valleys southwards from Brigg, the Wragby Till is essentially composed of Upper Jurassic detritus. The Heath Till, occurring west of Lincoln Edge—notably south-east of Gainsborough—and around Sleaford, reflects a change from Middle Jurassic through Lower Jurassic to Triassic derivation as it is traced westwards. In north-eastern Lincolnshire a channel, cut by sub-glacial drainage, starts near Kirmington and descends north-eastwards to more than 73 m below OD at Immingham, and appears to continue beneath the Basement Till of south-eastern Holderness. The fill of this channel is largely clay, some of which is stony and possibly till, but it also includes stoneless clay, sand and gravel. It is usually ascribed to the Anglian Stage by reference to the overlying Kirmington interglacial deposit (see below). During the final waning of the last pre-Ipswichian ice sheet, meltwater from the northern midlands deposited extensive sand and gravel along the River Trent, which at that time flowed north-eastwards from Newark and through the Lincoln Gap (Straw, 1963).

Interglacial stages

A Hoxnian age has been suggested for two small interglacial deposits in the region. The 'high-level' Speeton Shell Bed, occurring at about 30 m above OD in Filey Bay, consists of sand and silt containing estuarine mollusca. Although sparse pollen suggests equivalence with mixed oak forest assemblages known from the Ipswichian interglacial, the existence of overlying Basement Till indicates a Hoxnian age. At Kirmington, silts yielding estuarine mollusca,

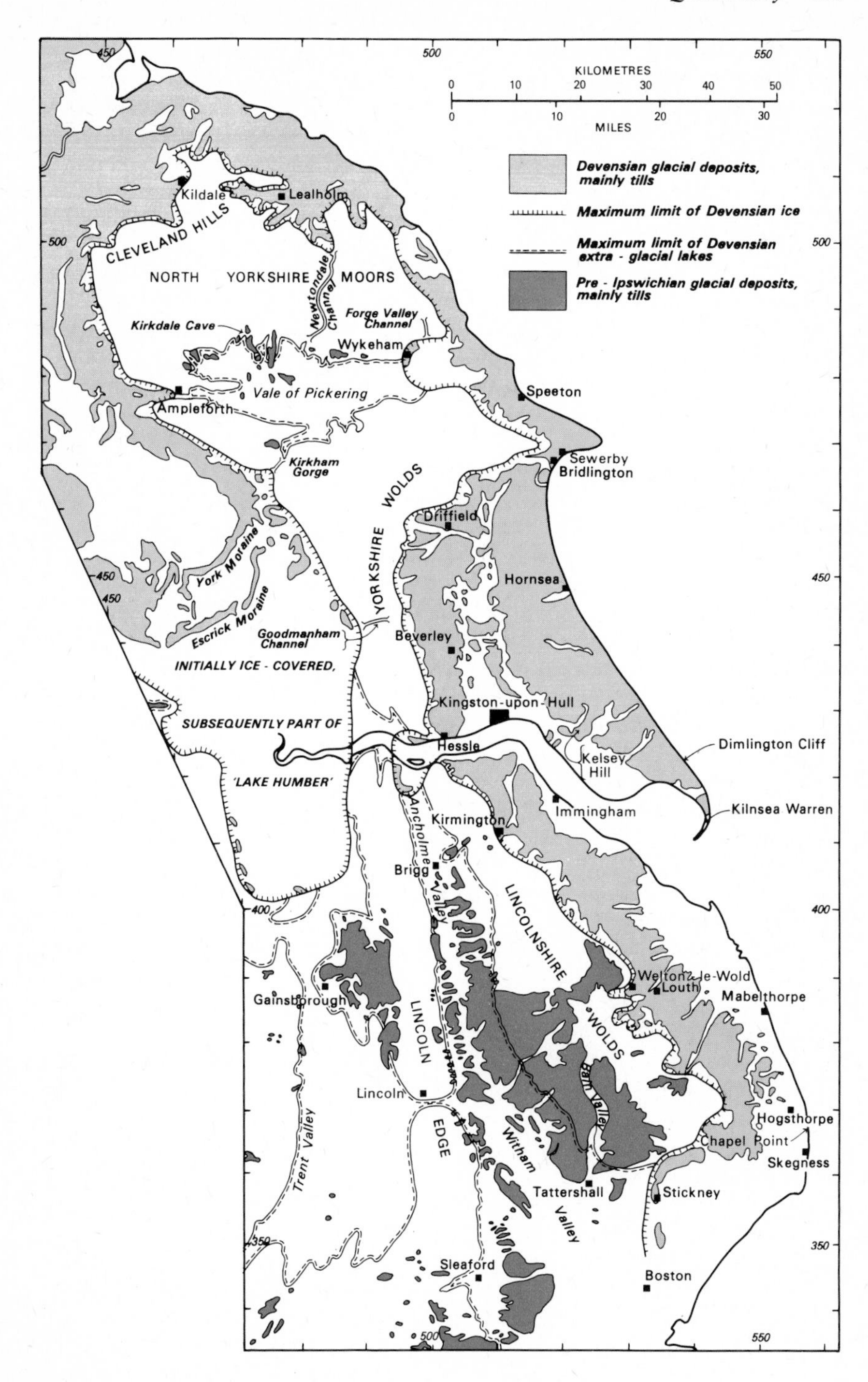

Figure 27 Generalised map of the glacial geology of the region

dinoflagellate cysts and plant debris, and containing a thin *Phragmites* peat near the base, rest at 19 to 25 m OD on the deposits occupying the sub-glacial channel described above. Pollen from the peat indicates a temperate forest environment, and has been considered to denote a Hoxnian age (Watts, 1959). A number of worked flint flakes, apparently derived from the overlying 'cannonshot gravel' which is probably a beach shingle, have Clactonian affinities, although one or two are more suggestive of the Acheulian (Boylan, 1966). Both typologies support a Hoxnian correlation. Recent excavations through Calcethorpe Till and a lower, locally developed, Welton Till at Welton-le-Wold (Alabaster and Straw, 1976) have revealed silts, sands and gravels which, although apparently periglacial deposits, have yielded a sparse interglacial mammalian fauna and a few Acheulian hand axes. Both the fauna and the implements suggest derivation from pre-existing Hoxnian deposits.

Deposits attributable to the Ipswichian interglacial have been found at a few localities in the region. Kirkdale Cave, formed along a bedding plane in Corallian limestone near Kirbymoorside and now largely quarried away, had a clay floor from which numerous hyena bones and teeth were collected; scattered remains of hippopotamus (virtually diagnostic of the Ipswichian), straight-tusked elephant (*Palaeoloxodon antiquus*), narrow-nosed rhinoceros (*Dicerorhinus hemitoechus*) and other mammals presumably represent dismembered pieces of carcases brought to the cave by hyenas. Ipswichian interglacial deposits in the Vale of York to the west show that sea level rose from at least 12 m below OD in the early part of the interglacial to a long-standing level at approximately 1 to 2 m above OD during the later part (Gaunt and others, 1974). This later level is reflected in Holderness by planation of the Basement Till and by formation of a now-buried Chalk cliff which runs inland from Sewerby via Driffield and Beverley to Hessle, whence it continues the length of Lincolnshire, passing near Keelby, Louth and Alford. At Sewerby (Figure 28) a shingle beach deposit banked against the old cliff has yielded mammalian remains, including the three species from Kirkdale Cave mentioned above, and marine and terrestrial mollusca. Possibly an equivalent deposit, the 'low-level' Speeton Shell Bed, occurs at present-day beach level near Reighton Gap. The molluscan fauna of this silt and sand, however, are reportedly similar to that in the 'high-level' shell bed (see above). In Lincolnshire, gravel pits near Tattershall expose peat between Wragby Till and dated Devensian deposits (see below). Pollen from the peat indicates an Ipswichian Zone IIb age and its insect fauna is that of a deciduous forest environment with summer temperatures warmer than today's (Girling, 1974).

Devensian Stage (glacial)

Although the Devensian commenced at least 70 000 years ago, glacial activity in the region was confined to only a few millenia and had ended by 13 000 years[1] ago at the latest; even then the ice invaded only peripheral areas. For much of the Devensian prior to the glacial phase the region experienced periglacial conditions of varying severity, producing ice-wedge pseudomorphs, involu-

[1] All dates of less than 50 000 years quoted in this chapter are in terms of radiocarbon years, which for geological purposes can be regarded as equivalent to true time.

tions, patterned ground and other evidence of cryoturbation. Head and aeolian deposits dating from this time occur, for example, at Sewerby, where they are banked against the interglacial cliff above the mammal-bearing shingle and are overlain by Devensian Skipsea (formerly Drab) Till (Figure 28). The low drainage base level, caused by world-wide eustatic lowering of sea

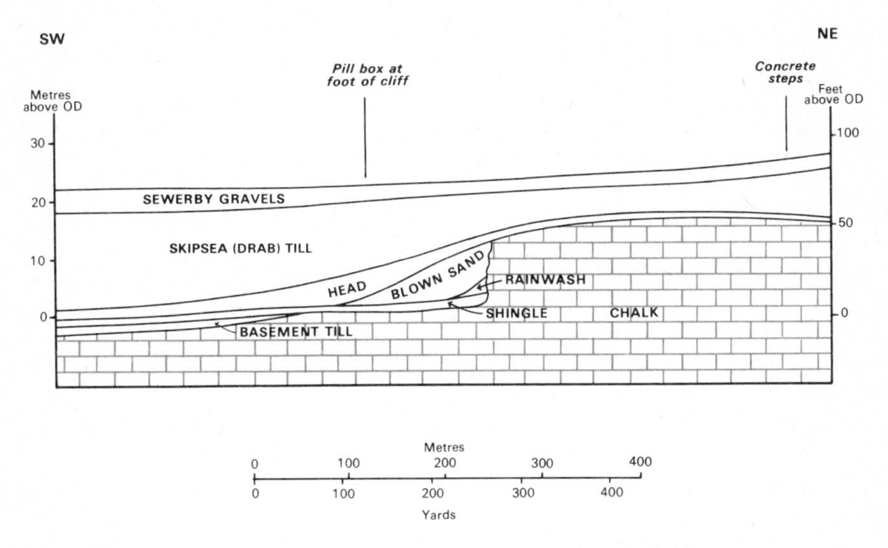

Figure 28 The Ipswichian cliff and beach shingle, Sewerby, Yorkshire

level during the Devensian, was responsible for most of the dissection which earlier Quaternary deposits have suffered. Fluvial sand and gravel exposed in quarries near Tattershall contain evidence of a short but marked mid-Devensian climatic amelioration (Girling, 1974). An organic silt bed within involuted gravel overlying the Ipswichian peat has yielded insects indicative of a high-latitude continental tundra environment, but a slightly higher bed contains a thermophilous insect fauna—some taxa of which are now confined to southern Europe—implying summer temperatures at least as warm as those of southern England today. Radiocarbon dates from these deposits and correlatives elsewhere in England show that this warm interval, the initial phase of the Upton Warren Interstadial Complex, occurred around 43 000 years ago. It was too brief for the climatic improvement to be reflected in the flora. The overlying sand and gravel in the Tattershall area are appreciably cryoturbated and contain numerous remains of mammoth (*Mammuthus primigenius*), woolly rhinoceros (*Coelodonta antiquatis*), bison, reindeer and other tundra-favouring mammals. Dimlington Cliff exposes another periglacial deposit; silt occupying shallow hollows in the top of the Basement Till, beneath Devensian Skipsea Till, contains moss and insects from a tundra environment. Radiocarbon dates slightly in excess of 18 000 years were obtained here (Penny and others, 1969).

During the Devensian glaciation ice encroached on the region from the west, north and east. A glacier derived from the Lake District crossed the Pennines by way of the Stainmore Gap and was split into two parts by the Cleveland Hills. The southern tongue was deflected southwards down the Vale of York, probably reaching the Isle of Axholme in north-western Lincolnshire (Gaunt, 1976) before quickly wasting back to a more prolonged front, where the crescentric till, sand and gravel ridges of the York and Escrick moraines were formed (Figure 27). These moraines curve north-eastwards and run into the hummocky tills, sands and gravels on the western sides of the Howardian and Hambleton hills. The northern tongue of Stainmore ice passed around the northern side of the Cleveland Hills where it was joined by Tyne valley and Cheviot–Scottish ice. The resulting deposits, mainly tills, are well seen in Robin Hood's Bay and Filey Bay (Plate 20). They thin out against the steep Cleveland slopes, but the Lealholm moraine and outcrops of till farther up Eskdale show that some ice entered the valleys. The Wykeham moraine demonstrates that ice entered the eastern end of the Vale of Pickering and blocked the drainage outlet permanently.

Farther south, North Sea ice moved across Holderness and into the Humber Gap, reaching almost to Brough and Winterton and effectively blocking drainage from the Vale of York and the Trent and Ancholme valleys. Two Devensian tills are clearly displayed in the Holderness cliffs. The lower, or Skipsea Till, overlies both the Ipswichian shingle and associated deposits at Sewerby (Figure 28), and the Late Devensian silts at Dimlington (Catt and Penny, 1966). It is a brownish grey clay with mainly Carboniferous erratics, some at least of Pennine origin, and an appreciable number of chalk fragments. The Skipsea is the more extensive of the two tills, lapping up the Chalk slopes and intruding into the Humber Gap. Exposures of the overlying Withernsea (formerly Purple) Till are limited to the cliffs south of Hornsea, and this deposit appears to be confined to south-eastern Holderness. It is a grey clay with a pinkish tinge that apparently owes its origin to the incorporation of Triassic materials. It contains a variety of northern English erratics but rather less chalk than the Skipsea Till. Both tills have a few Norwegian erratics. Pressure from Scandinavian ice may be responsible for the stone orientations of the tills in Holderness, which imply south-westerly movement (Penny and Catt, 1967). It should be noted here that the so-called Hessle Till of Holderness, previously regarded as a distinct unit overlying both Skipsea and Withernsea tills, has now been recognised to be the weathered derivative of these tills (Madgett and Catt, 1978).

The topography of Holderness provides evidence of relatively recent glacial deposition; subdued drumlinoid mounds, moraine-like ridges, kamiform sand deposits and numerous kettle holes are all present. At Kelsey Hill a sinuous ridge of sand and gravel running south-south-westwards, but now almost quarried away, rests on Skipsea Till and is partly overlain by what is probably weathered Withernsea Till. The most notable feature of the gravels is their fossil content, which includes numerous marine molluscs, mostly of temperate species, and the fresh-water bivalve *Corbicula fluminalis*, which lived in England during the Ipswichian interglacial but is now restricted to a few rivers in North Africa and the Middle East. Also present are the remains of mammals, including both temperate interglacial taxa and 'tundra' taxa. Since the general form of the Kelsey Hill deposit suggests an esker, it is assumed that the fossils were derived from pre-existing deposits and laid down by subglacial meltwater.

Devensian tills and sand and gravel are widespread in Lincolnshire east of the Wolds (Figure 27). Morainic features, meltwater deposits and drainage channels suggest that two successive ice fronts were established (Straw, 1961b). The earlier, maximum, limit is traceable from Winteringham and Horkstow at the mouth of the Ancholme Valley, along the eastern slopes of the Wolds past Kirmington, where till rests on the interglacial deposits, and around the southern end of the Wolds where, near Stickney, a subdued ridge of till suggests a continuation southwards beneath the recent deposits of the fenland. The later, younger, ice front runs from Immingham south-eastwards past Louth and disappears out to sea near Hogsthorpe.

Meltwater channels and associated deposits occur around the western, northern and eastern sides of the Cleveland Hills and North Yorkshire Moors, and are particularly distinct in Eskdale. Kendall, in his classic work of 1902, regarded these as ice-marginal phenomena formed subaerially, but some are now recognised as being entirely or partly sub-glacial (Gregory, 1965). However, the magnificent Newton Dale Channel was certainly utilised by meltwater flowing subaerially from the Esk Valley and its tributaries into the Vale of Pickering, and Forge Valley, west of Scarborough, was also, at least in its later stages, a subaerial spillway. Blockage of both the western and eastern ends of the Vale of Pickering produced a lake in which clays, and littoral and deltaic sands and gravels, were deposited. Water from Lake Pickering overflowed through the Kirkham Gorge and subsequently the River Derwent has continued to drain the vale by this route. Farther south, meltwater channels cut in the eastern slopes of the Yorkshire Wolds converge on the channel through the Wolds near Goodmanham which provided an outlet to the west. Blocking of the Humber Gap produced a lake occupying much of the Vale of York, into which the Derwent, Goodmanham and other channels discharged. This Lake Humber initially reached an elevation of 33 m above OD on the evidence of littoral deposits around its margins, but it soon fell to 10 to 14 m OD (Gaunt, 1974), and in this long-standing lake up to 20 m of laminated clay were deposited. Lake Humber extended up the Trent Valley and overflowed through the Lincoln Gap. Another lake, impounded by ice along the northern coast of Norfolk, stretched southwards across the fenland.

A radiocarbon date on organic deposits in a kettle hole near Kildale, on the western slopes of the Cleveland Hills, suggests that here the ice was waning by about 16 000 years ago (Jones, 1977). Dates from kettle holes in Holderness imply that ice had disappeared from that area by about 13 000 years ago. Pollen evidence from the Vale of York suggests clearance of ice from the Escrick moraine before about 12 400 years ago, and Lake Humber had certainly drained by 1300 years later (Gaunt and others, 1971). Once the effective drainage base level fell to late Devensian sea level, the rivers began to incise their courses to considerable depths below OD. Organic deposits from kettle holes and old lakes, including the Holderness 'meres', show that the region experienced both the temporary climatic amelioration now known as the Windermere Interstadial and, in the last millenium before the end of the Devensian 10 000 years ago, the severe climatic deterioration commonly referred to as the 'Younger Dryas' period. During the latter the climate was cold enough for cryoturbation structures to form and also sufficiently dry for blown sand to accumulate extensively, notably in north-western Lincolnshire and along the eastern margin of the Vale of York.

Flandrian Stage (post-glacial)

The world-wide eustatic rise in sea level during the Flandrian Stage resulted in the deposition of clays, silts, sands and gravels along the coasts and in the estuaries and valleys. Organic remains, including plants and mollusca, enable the succession of environments to be dated and related to the rising sea level (Gaunt and Tooley, 1974). Fresh-water peat, containing pollen from surrounding deciduous woodlands, formed about 6900 years ago, and lies at about 10 m below OD under Kingston upon Hull. A similar deposit, nearly 6700 years old, lies at about 9 m below OD beneath Immingham. The deposits are overlain by marine sands and silts. For a time the marine transgression appears to have been exceptionally rapid, for just south of Kilnsea Warren peat overlying marine clays at only 2.4 m below OD yielded a date of 6200 years. Younger deposits in the Ancholme Valley, the Humber Estuary and at Chapel Point on the Lincolnshire coast show that the transgression continued at a slower rate and with minor regressions until historic times.

Extensive growth of blanket bog peat, resulting from the rise in sea level and the onset of an oceanic climate, occurred in low-lying ill-drained areas during later Flandrian time (Smith, 1958). These peat deposits are most extensive in the Fens and in the Witham, Ancholme and Trent valleys. With the destruction of the natural forest cover in historic times, blown sand has accumulated in places, notably along the Lincolnshire coast, and the Devensian blown sands of north-western Lincolnshire have been partially redistributed.

Coastal changes similar to those that can be witnessed at present have probably been continuing for much of Flandrian time. The annual rate of erosion along the Holderness coast increases southwards from less than a metre near Bridlington to nearly 2.8 m near Kilnsea. This retreat of the cliff line increasingly exposes the shingle and sand spit of Spurn Point which, as historical evidence shows, has been destroyed and re-formed, progressively farther west, every 250 years or so (De Boer, 1964). In the Humber Estuary some land has been lost by marine transgression and channel migration, but other areas such as Broomfleet, Read's and Sunk islands have been reclaimed.

Peat and 'submerged forest' resting on till exposed near low tide mark along parts of the Lincolnshire coast have yielded Neolithic implements and show that sea level was still relatively low prior to about 4000 years ago. Subsequent marine transgression is indicated by the overlying soft grey clays, deposited within the upper part of the contemporaneous intertidal range, but in sheltered waters suggesting the existence of a protecting offshore bank or barrier. A higher peat layer, with evidence of salt workings and associated late Bronze Age and early Iron Age pottery, now between tide marks, reflects a later regression, as does the Roman settlement at Ingoldmells Point. These sites were subsequently buried during a more recent transgression by silt and clay containing brackish and marine faunas. The postulated offshore bank may have been destroyed in the late thirteenth and early fourteenth centuries when the Lincoln Marsh suffered severe inundations. At present the Lincolnshire coast varies in its reaction to tidal influences. Between Mablethorpe and Skegness it is undergoing slight erosion, but to the north and south depositional activity is dominant and the coastline is slowly advancing. Nowhere is this more obvious than at the medieval port of Boston, now some distance from the sea.

13. Economic minerals and water supply

The region has long been an important source of bulk minerals—in particular of sand, gravel, bedded ironstones and limestones. Working of ironstone is now limited to the Scunthorpe area, but sand, gravel and limestone remain important, and potash has recently been added to the list. For more than a century the quantities of ironstone and other bulk minerals extracted annually in the region have been measured in millions of tonnes.

The region also includes the more easterly onshore oil and gasfields, but these are small by comparison with the newer discoveries of the North Sea.

Among minor minerals, beds of phosphate nodules occur in the Cretaceous but are well below the economic feasibility level. Jet was formerly extensively mined in the Whitby area. The manufacture of alum reached a maximum of some 6000 tonnes per annum around 1769 but the industry has been extinct since the early years of the last century (see p. 40).

Energy sources

Coal In the last hundred years the development of the Yorkshire and East Midlands Coalfield has moved progressively eastwards beneath the Permian and Triassic cover, and the National Coal Board are currently planning a major colliery in the Selby–York area. Coal Measures extend beneath south-east Yorkshire and east Lincolnshire, but are much too deep for conventional mining of the coals they contain.

Jurassic coal seams were formerly of local importance in the Cleveland Hills (Danby, Eskdale and Coxwold), where there are widespread traces of shallow workings. The coals are thin and of poor quality, and were used mainly for lime burning.

Oil The first oil found in Lincolnshire was in the Carboniferous Limestone, at Nocton, but only a small quantity was produced. Subsequently oil, with variable amounts of associated gas, was discovered in Upper Carboniferous sandstones at Gainsborough, Corringham and Glentworth, where it is emplaced in a combination of structural and stratigraphical traps. Total production from these fields exceeds 360 000 tonnes.

Natural gas The search for gas in the Permian rocks of Yorkshire was initiated in 1938 on the evidence of Zechstein gas in Germany, with the Eskdale Dome as the first objective. A gasfield of modest size resulted from this project (although the incidental discovery of the Yorkshire potash field in the second Eskdale well was economically more significant) and this has been followed by the drilling of more than 40 exploration holes for gas elsewhere in the Permian basin north of the Humber. Results have been disappointing, with only one other small discovery, at Lockton, near Scarborough, also in Permian dolomites.

Exploration for gas in the North Sea was not encouraged by the scale of these early results, but was later inaugurated following a very large gas discovery in the Lower Permian (Rotliegendes sands) at Gröningen in the Netherlands. The largest offshore discoveries have been made in very thick Rotliegendes sands east of Norfolk; the smaller West Sole and Rough gas fields lie in thinner developments east of the Humber. The gas originates from the Coal Measures, being released as progressive subsidence of the basin increased the temperatures and pressure on these rocks; its survival and concentration in economic quantities depended on the availability of good reservoirs (usually the Basal Permian Sands) and an adequate cap-rock (particularly the Zechstein evaporites) to trap the migrating gas.

The small size of gas reserves found in the wells drilled in the Cleveland area and offshore near the coast of Yorkshire and Lincolnshire may be partly due to the relatively small amount of Coal Measures source rocks surviving in these areas. This reduces but does not eliminate the prospects, and the search is continuing.

Oil shales Oil shales provide a potential fuel source for the future. In Britain Lower Carboniferous shales have been exploited in Scotland and are now largely exhausted, but those of the Mesozoic of England remain to be developed when economic factors permit.

Bituminous shales of potential economic interest, interbedded with normal mudstones, are known in the Kimmeridge Clay of Lincolnshire and Yorkshire. The section penetrated at Donington-on-Bain, where there are large tonnages in the *eudoxus* and higher zones, is shown in Figure 17. In Yorkshire isolated occurrences are known at the first of these levels at Marton in the Vale of Pickering.

There are other bituminous shale horizons in the Jurassic rocks; of these the Upper Lias Jet Rock of east Yorkshire may justify further investigation, although there would be serious environmental problems in any large-scale exploitation.

Geothermal energy The scope for developing geothermal energy in the region is small, but hot water has occasionally been encountered in boreholes and there is a possibility of locating low-enthalpy sources. Exploitation would depend on finding a major aquifer connected vertically with hotter rocks at depth. The Carboniferous Limestone beneath the thinner Mesozoic cover on a structural high in mid-Lincolnshire or in the Howardian Hills fault belt might provide such a source, but the economics of locating and developing any such deep reservoir would be problematical.

Ironstone

There are bedded iron-rich rocks at many horizons, and four of these have been economically exploited—the Lower Lias Frodingham Ironstone, the Middle Lias Marlstone Rock and Cleveland Ironstone, the Northampton Sand Ironstone and the Dogger with the locally worked Rosedale 'magnetic ore', and the Lower Cretaceous Claxby Ironstone. Five other horizons locally approach ore grade—the *Pecten* Ironstone of north Lincolnshire, an oolitic ironstone development in the Eller Beck Formation in the Glaisdale district, the Cornbrash of Newton Dale, the Kellaways Rock of north Yorkshire and the Roach Stone ironstone of the Lincolnshire Wolds. These have not been exploited on a significant scale and the reserves are small.

Only the Frodingham Ironstone is now exploited. Working of the others ended because of exhaustion of the better quality ore (as in the Middle Lias) or in favour of much higher grade imported ironstone. Major tonnages of ore remain in the ground in some areas, particularly east of Scunthorpe and in the Northampton Sand Ironstone field.

The Frodingham Ironstone of north-west Lincolnshire is worked by opencast methods (Plate 3.2) over a distance of 12 km along its outcrop and is also mined to the east as far as Appleby (Figure 29). The town of Scunthorpe owes its rise since about 1860 to the resulting iron and steel industry. The average worked thickness of this chamositic and sideritic oolite, weathering to limonite, is about 9 m. It is a low-grade self-fluxing ore with an average iron content of 17 to 35 per cent. The ironstone has been proved by boreholes to extend eastwards beneath the northern Lincolnshire Wolds to the coast, and for some miles offshore. Annual production peaked at 5.6 million tonnes in the years 1961 and 1962.

The Middle Lias formerly carried the great ironstone field of Cleveland and a small field in south-west Lincolnshire. In Cleveland there are four seams of chamosite mudstone which, when traced southwards, all attenuate and are split by bands of shale. The Main Seam extends from west to east across the field and ranges in thickness from about 3.4 m in the north to under 2 m in the south of the worked area. It yielded practically the whole of the output of iron ore raised in Cleveland, which amounted to about 2 million tonnes in 1939. The iron content of the Main Seam reaches 30 per cent, but the lime content is low (5 per cent), and limestone had to be added as flux. The ore of the three other minor seams is of inferior quality but was worked in various localities. Production from the Cleveland Ironstone fluctuated around six million tonnes per annum from 1874 to 1913 and then declined steadily; working ceased in 1964.

In south-west Lincolnshire the upper 2.7 to 5 m of the Marlstone Rock has been worked opencast for iron ore. The best ore is a green, densely oolitic limestone in its unweathered state, yielding from 23 to about 28 per cent iron. It weathers to brown limonite at outcrop. This ironstone has been worked as far north as Belton but fails at Leadenham, 23 km S of Lincoln. The main area of exploitation was on the Leicestershire border and the last working mines were near Harlaxton, just beyond the boundary of the region.

In Lincolnshire the Northampton Sand Ironstone has been worked in places as far north as Lincoln, beyond which it is represented by ferruginous sands with ironstone concretions. Where fully developed the ironstone is a green oolitic carbonate more or less oxidised at outcrop to 'boxstone' limonite yielding from 31 to 38 per cent iron. Usually the uppermost 3 to 4.5 m of ore were worked, leaving the poor quality lower beds. The Dogger in north Yorkshire is also ferruginous and locally was sufficiently rich to be worked. A considerable quantity of ore was obtained from this horizon in Rosedale and on the coast around Staithes.

The Cretaceous Claxby Ironstone crops out along the western escarpment of the Lincolnshire Wolds southwards from Audleby, and was mined between Nettleton and Normanby-le-Wold. It is an oolitic clay ironstone with a

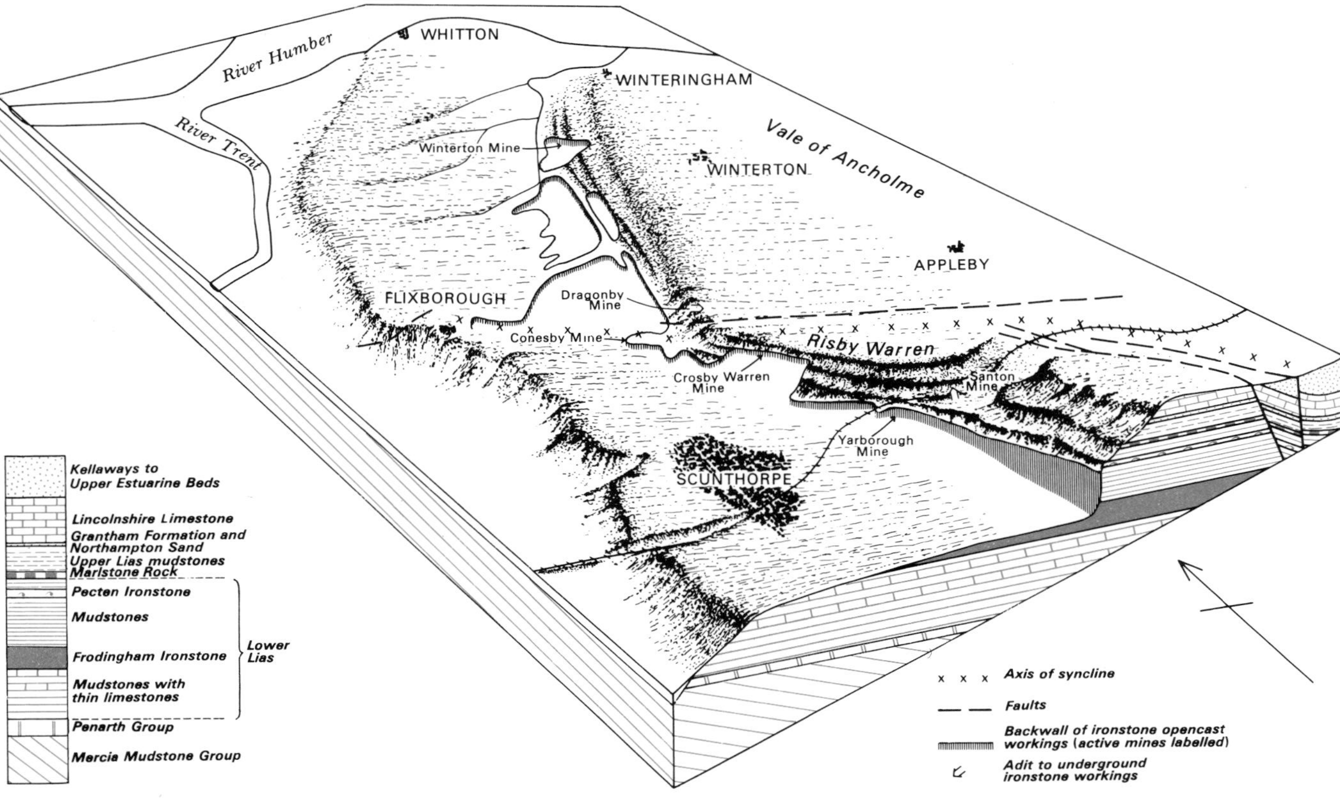

Figure 29 Block diagram of the Scunthorpe area showing the Frodingham Ironstone workings

maximum thickness of about 5 m though on average only about 2 m were workable ore, yielding up to 31 per cent iron. It is a soft ore which required special treatment before smelting.

Production figures are available in the official Mineral Statistics, from which the following totals are extracted:

Cleveland Ironstone (and minor Yorkshire sources including Rosedale)	371 363 000 tonnes
Frodingham Ironstone (up to 1974)	251 652 000 tonnes
Marlstone Rock, Northampton Sand Ironstone and Claxby Ironstone (minor) of Lincolnshire	104 891 000 tonnes

Other metals—geochemical surveys

The region has been included in a country-wide reconnaissance survey of geochemical anomalies based on stream sampling, carried out by Imperial College for the Wolfson Foundation. This has measured concentrations of a considerable range of elements—Al, As, Ba, Cd, Ca, Cr, Co, Cu, Ga, Fe, Pb, Li, Mg, Mo, Ni, Sc, Si, Sr, Sn, Va and Zn.

In a region underlain by Mesozoic sediments, largely masked by Drift and with a virtual absence of mineralisation, it is not to be expected that many notable anomalies will occur, and in fact most of the geochemical maps are almost entirely featureless. There are, however, areas with very high Mo. Molybdenum is known to be present in above-average quantities in some dark shales of euxinic type in the geological column, and Mo anomalies are indeed found in west Lincolnshire on part of the Lias outcrop (perhaps emanating from the *angulata* Zone shales) and on the Kimmeridge Clay outcrop near Market Rasen. These anomalies are discontinuous since the outcrops are largely blanketed by Drift. They are of considerable potential importance in relation to livestock farming.

Some of the visible features are evidently related to industrial contamination, especially around Scunthorpe, where Pb, Zn, Ni as well as Mo are significantly increased. The molybdenum might relate to a Lias source, but the Pb and Zn are likely to be industrial contaminants, as is the Ni although its anomaly is not coincident with that of the Pb and Zn. The geochemical survey thus has value in monitoring pollution, independently of features which can be counted as of geological origin.

Evaporite minerals

Evaporites are widely associated with the Permian and Triassic rocks—calcium sulphate as gypsum or anhydrite extends through most of the region, halite occurs in east Yorkshire and beneath the North Sea, and there is potash also in the central part of the original Permian basin. Reserves of all the evaporite minerals are very large in relation to production, particularly if the reserves of the offshore area are taken into account, but exploitation involves major technical and economic problems.

Gypsum and anhydrite Gypsum occurs as multiple thin beds in the Mercia Mudstone of Nottinghamshire and west Lincolnshire. The thickest horizon, near the top of the group, is worked near Newark as a source of plaster, mainly used in the manufacture of plaster board, and was formerly exploited at

Gainsborough. The Upper Permian Billingham Main Anhydrite was formerly mined on the Tees at Billingham, for manufacture of sulphuric acid.

Salt (halite) Rock salt occurs in major thickness in the Upper Permian of east Yorkshire and Teesside; it is worked by solution mining in the Middlesbrough area. Thick salt is also present in the Lower Keuper of east Yorkshire, as well as offshore, but has not been exploited.

Potash Potassium and magnesium salts are present in Upper Permian rocks under east Yorkshire. The Fordon Evaporites contain polyhalite, but this has no commercial value. Sylvine occurs in the Boulby Halite and the Upper Halite, the horizons being known at the Boulby Potash and the Upper Potash respectively. The Boulby Potash (up to 45 per cent KCl) consists of lenticular bodies up to 11 m thick under Staithes and Whitby, but the Upper Potash (up to 26 per cent KCl) is a more consistent bed up to 8.5 m thick (including some halite-rich mudstone). Production by pillar and stall mining started in 1973 with a daily output of 7100 tonnes. The eventual capacity is estimated at one million tonnes per year. To the end of 1975 almost 60 000 tonnes of marketable KCl had been produced. Much is used for compound fertilizers but a small proportion is refined further for specialised industrial uses.

Bulk minerals

Building stone Many of the sandstones and limestones in the region are valuable as building stones. In Lincolnshire, the Marlstone Rock and the Ancaster Freestone are well known for this purpose, and the latter is still worked (Plate 6.1) for the construction of public buildings in London, Cambridge and elsewhere. The 'Silver Beds' of the Lincolnshire Limestone have been employed in Lincoln Cathedral. In the Howardian Hills the Dogger around Terrington was used in the twelfth century in the construction of Sheriff Hutton Castle. Whitby Abbey, Covent Garden, the foundations of the old Waterloo Bridge and London Bridge, and many other public buildings are constructed of sandstones of the Saltwick Formation from the Aislaby district. The Cave Oolite in the Humber area was used in building the Holderness monasteries and Hull Docks. The Kellaways Rock has been used in the Scarborough district, and the Lower Calcareous Grit in the Howardian Hills is of local importance. The fine white Hildenley Limestone was used in the interior work of Kirkham Abbey, near Malton.

In the Lincolnshire Wolds, around Salmonby, the Spilsby Sandstone is suitable for building purposes, while farther north the Tealby Limestone has been used around Tealby and Walesby, and Chalk was employed in the construction of Louth Abbey.

Lime and cement Jurassic limestones have been extensively burnt for lime production, principally for agricultural use. The principal sources are the Malton, Hambleton, Whitwell, and Cave oolites and the Lincolnshire Limestone. Local use was made of the Dogger around Stokesley, the Hydraulic Limestone in the Howardian Hills and the Grey Limestone of the Scarborough Formation in parts of the Cleveland Hills. Prior to the introduction of Portland cement, several Jurassic limestones containing suitable proportions of alumina and silica, such as those in the lower part of the Lower Lias, the Middle Jurassic Hydraulic Limestone and the North Grimston Cementstones, were used to produce a natural mixture for 'hydraulic' cement. Nodules from the Upper Lias and from the Speeton Clay have also been collected locally for the same purpose. Several Jurassic limestones have been used as fluxes, and at present small amounts of the Malton Oolite are taken for this purpose.

Although Kirton Cementstones from the Kirton area (Plate 6.2) are mixed with Liassic clay (Plate 5) to produce Portland cement, the principal source of this product is the Chalk, to which is added Upper Jurassic or Quaternary clays. The largest extraction is from the Chalk of the Humber area, which is also increasingly worked for whiting. Melton Chalk Pit is probably the largest whiting producer in Britain. Chalk for whiting is quarried also near Louth. Chalk is worked for agricultural usage, and in recent years has been used as bulk fill in road construction. The amount of Chalk extracted in the region for various purposes during 1973 is quoted as agricultural 177 000, industrial 684 000, road construction 336 000, cement and other uses 1 229 000 tonnes.

Roadstone Several of the Jurassic limestones have been quarried for road aggregate, the Hambleton and Malton oolites and the Kirton Cementstones still being worked for this purpose. The calcareous sandstones of the Brandsby Roadstone (Scarborough Formation) of the Howardian Hills and the siliceous Moor Grit of the Scalby Formation north of the Esk Valley also provide good quality aggregate. Although the Cleveland Dyke was formerly quarried extensively, and also mined, for roadstone, it is now almost worked out, the only remaining production being near Great Ayton.

Bricks and clay The Keuper Marl at Bishop Wilton in east Yorkshire has been dug for brick-clay in the past. The Middle Lias clays around Lincoln, the Middle Jurassic shales on the Yorkshire coast, the Blisworth Clay in north Lincolnshire, the Kimmeridge Clay in Lincolnshire and the Upper Jurassic clays in the Vale of Pickering and the Howardian Hills have all been used in the brick and tile industry. Most of these sources are now disused, but thick deposits of boulder clay are actively worked in some localities. Transported Oxford Clay bricks from the south Midlands are now widely employed in the region.

Sand and gravel Gravels and sands are widely worked for concrete production and road construction, but on a smaller scale compared to adjacent regions. Sand is a component of mortar and plaster, and is also used for glass-making and moulding. The value of the deposits depends on the amount of impurities present, the nature of the pebbles in the gravels, and transport costs.

The more extensive workings of glacial sand and gravel are at the eastern end of the Vale of Pickering, in Holderness (notably near Brandesburton and Keyingham) and along the eastern side of the Lincolnshire Wolds. Beach gravels formed around Lake Humber have been worked between Pocklington and Brough, but the high proportion of chalk and limestone pebbles limits their use. Wet working of river gravels has long been carried out along former courses of the Trent through the Lincoln Gap, and along the existing Trent as far north as Torksey, beyond which the increasing depth of overburden currently renders them uneconomic. These river gravels are of good quality, the pebbles being mainly quartzites derived from the Trias. River gravels in the Bain Valley are also extracted by wet working; they contain appreciable proportions of chalk and flint pebbles. Sand is extracted from a low terrace of the Trent near Messingham and from aeolian deposits in the Scunthorpe and Nettleton areas, being used for moulding, making coloured glass and adjusting the lime/silica ratio of blast furnace burden. A silty glaci-lacustrine sand near North Kelsey is used for tile manufacture.

In recent years sand and gravel have been dredged from the Humber approaches; besides supplying the Humber, Tees and Tyne regions, some of this aggregate is exported to the continent. Beach deposits in Filey Bay were worked during the 1960s, but this was terminated for environmental reasons.

The Spilsby Sandstone near Nettleton is currently being used in plaster production. Friable siliceous sands from the Jurassic succession have been used for glass manufacture. The best quality came from the Scalby Formation at Hutton's Ambo, near Malton, but the reserves there are now limited. Other sources were the Kellaways Rock at Burythorpe, also near Malton, and the Kellaways Sand near South Cave and South Newbald. These and other Jurassic sands such as those in the Corallian succession of the Pickering area have also been used for refractory and moulding purposes.

Water supply

The major population centres and industrial complexes are supplied largely by groundwater, but increasing use is being made of river abstraction. Many villages are sited where springs issue from the Chalk or from Jurassic limestones and sandstones, and some supplies in country areas are still obtained from these sources.

The Chalk is the most important aquifer, supplying up to 180, 184 and 5 megalitres per day* respectively to Yorkshire, north and south Lincolnshire. Although hard, the water is suitable for domestic and most industrial purposes, but overpumping can result in upward 'coning' of saline water from lower levels and saline intrusion in coastal areas. The rising nitrate content of some Chalk and Lincolnshire Limestone groundwater is causing concern. Scarborough and the Vale of Pickering receive much of their water from boreholes into Corallian rocks, which are partly charged by swallow holes in the upper Derwent Valley; 20 megalitres per day are drawn from boreholes at Irton and Osgodby. Industry in the North Ferriby area takes water from Upper Jurassic limestones and sandstones, and from the Cave Oolite. The Spilsby Sandstone supplies up to 25 megalitres per day to south-east Lincolnshire, including Boston, Skegness and Mablethorpe, while much of north and central Lincolnshire, including Scunthorpe, Gainsborough, Lincoln and Grantham, obtains water from boreholes into the Lincolnshire Limestone. Some supplies of gypsiferous water for industry, notably brewing at Newark, have been taken from the underlying Triassic rocks.

Water is taken from the River Esk to supply Whitby and some local villages, while Scarborough receives additional supplies from springs issuing from Corallian rocks in Cayton Bay. Kingston upon Hull augments its supplies

* 1 megalitre = 1000 cubic metres = 1000 tonnes = 220 000 gallons.

from the Chalk with up to 68 megalitres per day from the River Hull, and will receive up to 77 megalitres per day from the lower reaches of the River Derwent when the Barmby Barrage Scheme is completed. When in full operation, the Trent–Witham–Ancholme Water Transfer Scheme and the Great Eau Scheme will supply up to 57 and 73 megalitres per day respectively to north Lincolnshire. Water from the Trent and Ancholme is used by industry around Scunthorpe; Grantham still takes water from the upper reaches of the Witham and Boston takes some from the Bain.

A few western parts of the region receive water from sources farther west. Stokesley obtains supplies from Teesdale, and both Lincoln and Gainsborough augment their supplies with water from boreholes into Triassic sandstones in the East Retford area.

14. Select bibliography

Geological Survey memoirs

These memoirs are published by Her Majesty's Stationery Office, London. Some of the older works are out of print.

General memoirs

FOX-STRANGWAYS, C. 1892. The Jurassic Rocks of Britain. Vols. 1 and 2 [Yorkshire].

JUKES-BROWNE, A. J. 1900–04. The Cretaceous Rocks of Britain. [3 vols.].

WOODWARD, H. B. 1893–95. The Jurassic Rocks of Britain. Vols. 3–5 [Yorkshire excepted].

Regional and sheet memoirs

BARROW, G. 1888. The geology of North Cleveland.

DAKYNS, J. R. and FOX-STRANGWAYS, C. 1885. The geology of Bridlington Bay.

— — 1886. The geology of the country around Driffield.

— — and CAMERON, A. G. 1886. The geology of the country between York and Hull.

EDWARDS, W. 1967. Geology of the country around Ollerton. 2nd Edition.

FOX-STRANGWAYS, C. 1881. The geology of the Oolitic and Liassic Rocks to the north and west of Malton.

— 1884. The geology of the country north-east of York and south of Malton.

— 1885. The geology of Eskdale, Rosedale, &c.

— 1904. The geology of the Oolitic and Cretaceous Rocks south of Scarborough. 2nd Edition.

— and BARROW, G. 1915. The geology of the country between Whitby and Scarborough. 2nd Edition.

— CAMERON, A. G. and BARROW, G. 1886. The geology of the country around Northallerton and Thirsk.

JUKES-BROWNE, A. J. 1885. The geology of the south-west part of Lincolnshire, with parts of Leicestershire and Nottinghamshire.

— 1887. The geology of east Lincolnshire, including the country near the towns of Louth, Alford and Spilsby.

REID, C. 1885. The geology of Holderness and the adjoining parts of Yorkshire and Lincolnshire.

SKERTCHLY, S. B. J. 1877. Geology of the Fenland.

SMITH, E. G., RHYS, G. H. and GOOSSENS, R. F. 1973. Geology of the country around East Retford, Worksop and Gainsborough.

USSHER, W. A. E. 1890. The geology of parts of north Lincolnshire and south Yorkshire.

— JUKES-BROWNE, A. J. and STRAHAN, A. 1888. The geology of the country around Lincoln.

Economic and water supply memoirs

FOX-STRANGWAYS, C. 1906. The water supply (from underground sources) of the East Riding of Yorkshire, together with the neighbouring portions of the vales of York and Pickering: with records of sinkings and borings.

HOLLINGWORTH, S. E. and TAYLOR, J. H. 1951. The Mesozoic Ironstones of England. The Northampton Sand Ironstone: stratigraphy, structure and reserves.
TAYLOR, J. H. 1949. Petrology of the Northampton Sand Ironstone Formation.
WHITEHEAD, T. H., ANDERSON, W., WILSON, V. and WRAY, D. A. 1952. The Mesozoic Ironstones of England. The Liassic Ironstones.
WOODWARD, H. B. 1904. The water supply of Lincolnshire from underground sources: with records of sinkings and borings.

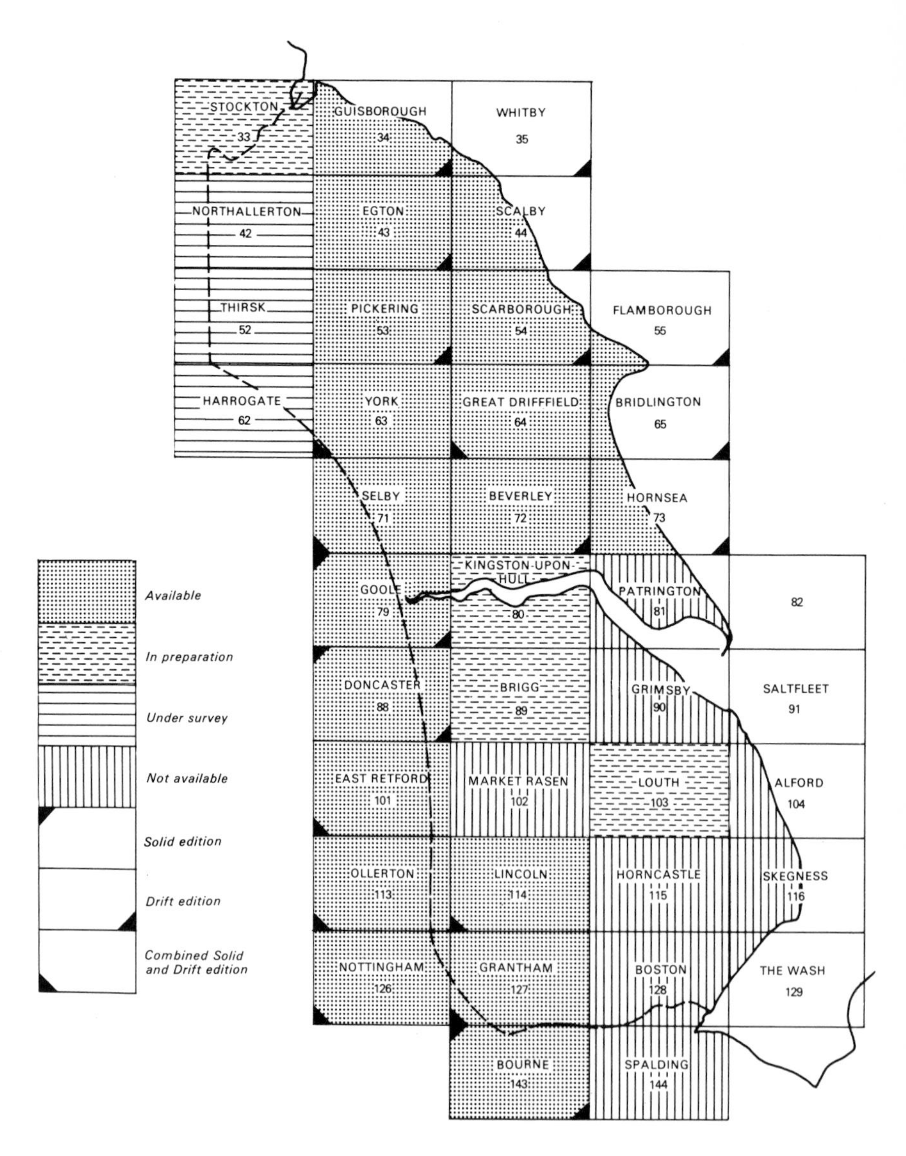

Figure 30 Availability index map of one-inch and 1 : 50,000 geological sheets in the region

Introduction (Chapter 1)

RAYNER, D. H. and HEMINGWAY, J. E. (Editors). 1974. *The geology and mineral resources of Yorkshire.* (Leeds: Yorkshire Geological Society.) ix + 405 pp.

SWINNERTON, H. H. and KENT, P. E. 1976. *The geology of Lincolnshire from the Humber to the Wash.* (2nd Edition, with revisions and additions by Sir Peter Kent.) (Lincoln: Lincolnshire Naturalists Union.) xiv + 130 pp.

WOODLAND, A. W. (Editor). 1975. *Petroleum and the continental shelf of north-west Europe.* Vol. 1, Geology. (Barking: Applied Science Publishers, for Institute of Petroleum.) ix + 501 pp.

Concealed pre-Permian rocks (Chapter 2)

HOWITT, F. and BRUNSTROM, R. G. W. 1966. The continuation of the East Midlands Coal Measures into Lincolnshire. *Proc. Yorkshire Geol. Soc.*, Vol. 35, pp. 549–564.

KENT, P. E. 1967. Outline geology of the Southern North Sea Basin. *Proc. Yorkshire Geol. Soc.*, Vol. 36, pp. 1–22.

WILLS, L. J. 1973. A palaeogeological map of the Palaeozoic floor below the Permian and Mesozoic formations in England and Wales with inferred and speculative reconstructions of the Palaeozoic outcrops in adjacent areas in Permo-Triassic times. *Mem. Geol. Soc. London*, No. 7, 23 pp.

Permian (Chapter 3)

BRUNSTROM, R. G. W. and WALMSLEY, P. J. 1969. Permian evaporites in North Sea Basin. *Bull. Am. Assoc. Pet. Geol.*, Vol. 53, pp. 870–883.

KENT, P. E. 1967. *q.v.*

SMITH, D. B. 1974. Permian *in* RAYNER, D. H. and HEMINGWAY, J. E. (Editors), *q.v.*

TAYLOR, J. C. M. and COLTER, V. S. 1975. Zechstein of the English sector of the Southern North Sea Basin *in* WOODLAND, A. W. (Editor), *q.v.*

Triassic (Chapter 4)

BALCHIN, D. A. and RIDD, M. F. 1970. Correlation of the younger Triassic rocks across eastern England. *Q. J. Geol. Soc. London*, Vol. 126, pp. 91–101.

GAUNT, G. D., IVIMEY-COOK, H. C., PENN, I. E. and COX, B. M. 1980. Mesozoic rocks proved by I.G.S. boreholes in the Humber and Acklam areas. *Rep. Inst. Geol. Sci.*, No. 79/13, 34 pp.

KENT, P. E. 1955. The Market Weighton Structure. *Proc. Yorkshire Geol. Soc.*, Vol. 30, pp. 197–227.

WARRINGTON, G. 1974. Trias *in* RAYNER, D. H. and HEMINGWAY, J. E. (Editors), *q.v.*

Jurassic (Chapters 5–7)

ARKELL, W. J. 1933. *The Jurassic System in Great Britain.* (Oxford: Clarendon Press.) viii + 681 pp.

BAIRSTOW, L. 1969. Lower Lias in *William Smith Bicentenary Meeting, International Field Symposium on the British Jurassic.* HEMINGWAY, J. E., WRIGHT, J. K. and TORRENS, H. S. (Editors). Excursion No. 3, Guide for north-east Yorkshire. (Keele: Keele University Geology Department.)

BATE, R. H. 1967. Stratigraphy and palaeogeography of the Yorkshire oolites and their relationship with the Lincolnshire Limestone. *Bull. Br. Mus. (Nat. Hist.), Geology*, Vol. 14, pp. 111–141.

BLACK, M. 1928. "Washouts" in the Estuarine Series of Yorkshire. *Geol. Mag.*, Vol. 65, pp. 301–307.

— 1929. Drifted plant-beds of the Upper Estuarine Series of Yorkshire. *Q. J. Geol. Soc. London*, Vol. 85, pp. 389–437.

— 1934. Sedimentation of the Aalenian rocks of Yorkshire. *Proc. Yorkshire Geol. Soc.*, Vol. 22, pp. 265–279.

CALLOMON, J. H. 1968. The Kellaways Beds and the Oxford Clay in *The Geology of the East Midlands*. SYLVESTER-BRADLEY, P. C. and FORD, T. D. (Editors). (Leicester: Leicester University Press.) xx + 400 pp.

CASEY, R. 1962. The ammonites of the Spilsby Sandstone, and the Jurassic–Cretaceous boundary. *Proc. Geol. Soc. London*, No. 1598, pp. 95–100.

COPE, J. C. W. 1974. New information on the Kimmeridge Clay of Yorkshire. *Proc. Geol. Assoc.*, Vol. 85, pp. 211–221.

DINGLE, R. V. 1971. A marine geological survey off the north-east coast of England (western North Sea). *J. Geol. Soc. London*, Vol. 127, pp. 303–338.

GALLOIS, R. W. 1974. Geological report on the inland site investigation 1972. Pp. 27–166 in *The Wash Water Storage Scheme: Report on the geological investigations for the feasibility study, March 1972.* (Institute of Geological Sciences, unpublished report.) v + 166 pp.

GAUNT, G. D. and OTHERS. 1980 *q.v.*

HARRIS, T. M. 1953. The geology of the Yorkshire Jurassic flora. *Proc. Yorkshire Geol. Soc.*, Vol. 29, pp. 63–71.

HEMINGWAY, J. E. 1949. A revised terminology and subdivision of the Middle Jurassic rocks of Yorkshire. *Geol. Mag.*, Vol. 86, pp. 67–71.

— 1974. Jurassic *in* RAYNER, D. H. and HEMINGWAY, J. E. (Editors), *q.v.*

— and KNOX, R. W. O'B. 1973. Lithostratigraphical nomenclature of the Middle Jurassic strata of the Yorkshire Basin of north-east England. *Proc. Yorkshire Geol. Soc.*, Vol. 39, pp. 527–535.

HEYBROEK, P. 1975. On the structure of the Dutch part of the Central North Sea Graben *in* WOODLAND, A. W. (Editor), *q.v.*

HOWARTH, M. K. 1973. The stratigraphy and ammonite fauna of the Upper Liassic Grey Shales of the Yorkshire coast. *Bull. Br. Mus. (Nat. Hist.), Geology*, Vol. 24, pp. 237–277.

KENT, P. E. 1975. The Grantham Formation of the East Midlands: revision of the Middle Jurassic, Lower Estuarine Beds. *Mercian Geol.*, Vol. 5, pp. 305–327.

PARSONS, C. F. 1977. A stratigraphic revision of the Scarborough Formation. *Proc. Yorkshire Geol. Soc.*, Vol. 41, pp. 203–222.

SENIOR, J. R. and EARLAND-BENNETT, P. M. 1973. The Bajocian ammonite *Hyperlioceras rudidiscites* S. Buckman in eastern England and its significance. *Proc. Yorkshire Geol. Soc.*, Vol. 39, pp. 319–326.

SIMPSON, M. 1868. *A guide to the geology of the Yorkshire coast.* 4th Edition. (Whitby: S. Reed.)

SMITH, W. 1832. *Stratification in Hackness Hills. (12 chains to one inch). Vertical sections of the strata.* (Publisher: W. Day, Lithographer to the King, 17 Gate St., Lincolns Inn Fields.)

SYLVESTER-BRADLEY, P. C. 1953. A stratigraphical guide to the fossil localities of the Scarborough district. Pp. 19–48 in *The natural history of the Scarborough district*, Vol. 1. (Scarborough: Scarborough Field Naturalists Society.) xii + 296 pp.

TATE, R. and BLAKE, J. F. 1876. *The Yorkshire Lias.* (London: John van Voorst.) iv + 475 pp.

WILSON, V. 1948. East Yorkshire and Lincolnshire. *Br. Reg. Geol.* (London: HMSO.)
WRIGHT, C. D. 1976. New outcrops of Ampthill Clay north of Market Weighton, north Yorkshire, and their structural implications. *Proc. Yorkshire Geol. Soc.*, Vol. 41, pp. 127–140.
WRIGHT, J. K. 1968. The stratigraphy of the Callovian rocks between Newtondale and the Scarborough coast, Yorkshire. *Proc. Geol. Assoc.*, Vol. 79, pp. 363–399.
— 1972. The stratigraphy of the Yorkshire Corallian. *Proc. Yorkshire Geol. Soc.*, Vol. 39, pp. 225–266.

Cretaceous (Chapters 8–9)

BOWER, C. R. and FARMERY, J. R. 1910. The zones of the Lower Chalk of Lincolnshire. *Proc. Geol. Assoc.*, Vol. 21, pp. 333–359.
JEANS, C. V. 1968. The origin of the montmorillonite of the European Chalk with special reference to the Lower Chalk of England. *Clay Miner.*, Vol. 7, pp. 311–329.
— 1973. The Market Weighton structure: tectonics, sedimentation and diagenesis during the Cretaceous. *Proc. Yorkshire Geol. Soc.*, Vol. 39, pp. 409–444.
JEFFERIES, R. P. S. 1963. The stratigraphy of the *Actinocamax plenus* Subzone (Turonian) in the Anglo-Paris Basin. *Proc. Geol. Assoc.*, Vol. 74, pp. 1–33.
KENT, P. E. 1967. *q.v.*
NEALE, J. W. 1974. Cretaceous *in* RAYNER, D. H. and HEMINGWAY, J. E. (Editors), *q.v.*
RAWSON, P. F., CURRY, D., DILLEY, F. C., HANCOCK, J. M., KENNEDY, W. J., NEALE, J. W., WOOD, C. J. and WORSSAM, B. C. 1978. A correlation of Cretaceous rocks in the British Isles. *Spec. Rep. Geol. Soc. London*, No. 9, 70 pp.
ROWE, A. W. 1929. The zones of the White Chalk of Lincolnshire. *The Naturalist*, No. 875, pp. 411–439.
SMART, J. G. O. and WOOD, C. J. 1976. South Humberside *in* Field Meetings. *Proc. Yorkshire Geol. Soc.*, Vol. 40, pp. 586–593.
WOOD, C. J. and SMITH, E. G. 1978. Lithostratigraphical classification of the Chalk in North Yorkshire, Humberside and Lincolnshire. *Proc. Yorkshire Geol. Soc.*, Vol. 42, pp. 263–287.
WRIGHT, C. W. 1935. The Chalk Rock fauna in east Yorkshire. *Geol. Mag.*, Vol. 72, pp. 441–442.
— and WRIGHT, E. V. 1942. The Chalk of the Yorkshire Wolds. *Proc. Geol. Assoc.*, Vol. 53, pp. 112–127.

Tertiary (Chapter 10)

DINGLE, R. V. 1971. *q.v.*

Structure (Chapter 11)

BISAT, W. S., PENNY, L. F. and NEALE, J. W. 1962. *Geology around the university towns: Hull.* Geologists' Association Guides, No. 11. (Colchester: Geologists' Association.)
BRUNSTROM, R. G. W. and WALMSLEY, P. J. 1969. *q.v.*
DINGLE, R. V. 1971. *q.v.*
HANCOCK, J. M. and SCHOLLE, P. A. 1975. Chalk of the North Sea *in* WOODLAND, A. W. (Editor), *q.v.*

HEMINGWAY, J. E. 1974. *q.v.*
HEYBROEK, P. 1975. *q.v.*
KENDALL, P. F. 1905. Sub-report on the concealed portion of the coalfield of Yorkshire, Derbyshire and Nottinghamshire. Appendix III, pp. 18–35 in Final report of the Royal Commission on coal supplies. (London: HMSO.)
KENT, P. E. 1955. *q.v.*
— 1974. Structural history *in* RAYNER, D. H. and HEMINGWAY, J. E. (Editors), *q.v.*
LEES, G. M. and COX, P. T. 1937. The geological basis of the present search for oil in Great Britain by the D'arcy Exploration Company, Limited. *Q. J. Geol. Soc. London*, Vol. 93, pp. 156–194.
MARIE, J. P. P. 1975. Rotliegendes stratigraphy and diagenesis *in* WOODLAND, A. W. (Editor), *q.v.*
SWINNERTON, H. H. and KENT, P. E. 1976. *q.v.*
VERSEY, H. C. 1923. Note on the stratigraphy of the Chalk in Yorkshire. *Trans. Leeds Geol. Assoc.*, Part 19, pp. 34–37.
ZIEGLER, P. A. 1975. North Sea Basin history in the tectonic framework of north-western Europe *in* WOODLAND, A. W. (Editor), *q.v.*

Quaternary (Chapter 12)

ALABASTER, C. and STRAW, A. 1976. The Pleistocene context of faunal remains and artifacts discovered at Welton-le-Wold, Lincolnshire. *Proc. Yorkshire Geol. Soc.*, Vol. 41, pp. 75–94.
BOYLAN, P. J. 1966. The Pleistocene deposits of Kirmington, Lincolnshire. *Mercian Geol.*, Vol. 1, pp. 339–350.
CATT, J. A. and PENNY, L. F. 1966. The Pleistocene deposits of Holderness, east Yorkshire. *Proc. Yorkshire Geol. Soc.*, Vol. 35, pp. 375–420.
DE BOER, G. 1964. Spurn Head: its history and evolution. *Trans. Inst. Br. Geogr.*, Publ. No. 34, pp. 71–89.
GAUNT, G. D. 1974. A radiocarbon date relating to Lake Humber. *Proc. Yorkshire Geol. Soc.*, Vol. 40, pp. 195–197.
— 1976. The Devensian maximum ice limit in the Vale of York. *Proc. Yorkshire Geol. Soc.*, Vol. 38, pp. 631–637.
— BARTLEY, D. D. and HARLAND, R. 1974. Two interglacial deposits proved in boreholes in the southern part of the Vale of York and their bearing on contemporaneous sea levels. *Bull. Geol. Surv. G.B.*, No. 48, pp. 1–23.
— JARVIS, R. A. and MATTHEWS, B. 1971. The late Weichselian sequence in the Vale of York. *Proc. Yorkshire Geol. Soc.*, Vol. 38, pp. 281–284.
— and TOOLEY, M. J. 1974. Evidence for Flandrian sea-level changes in the Humber estuary and adjacent areas. *Bull. Geol. Surv. G.B.*, No. 48, pp. 25–41.
GIRLING, M. 1974. Evidence from Lincolnshire of the age and intensity of the mid-Devensian temperate episode. *Nature, London*, Vol. 250, p. 270.
GREGORY, K. J. 1965. Proglacial Lake Eskdale after sixty years. *Trans. Inst. Br. Geogr.*, Publ. No. 36, pp. 149–162.
JONES, R. L. 1977. Late Devensian deposits from Kildale, north-east Yorkshire. *Proc. Yorkshire Geol. Soc.*, Vol. 41, pp. 185–188.
KENDALL, P. F. 1902. A system of glacier-lakes in the Cleveland Hills. *Q. J. Geol. Soc. London*, Vol. 58, pp. 471–571.
LEWIN, J. 1969. The Yorkshire Wolds, a study in geomorphology. *Univ. Hull Occas. Pap. Geogr.*, No. 11, 89 pp.
MADGETT, P. A. and CATT, J. A. 1978. Petrography, stratigraphy and weathering of late Pleistocene tills in east Yorkshire, Lincolnshire and north Norfolk. *Proc. Yorkshire Geol. Soc.*, Vol. 42, pp. 55–108.

PENNY, L. F. 1974. Quaternary *in* RAYNER, D. H. and HEMINGWAY, J. E. (Editors), *q.v.*
— and CATT, J. A. 1967. Stone orientation and other structural features of tills in east Yorkshire. *Geol. Mag.*, Vol. 104, pp. 344–360.
— COOPE, G. R. and CATT, J. A. 1969. Age and insect fauna of the Dimlington Silts, east Yorkshire. *Nature, London*, Vol. 224, pp. 65–67.
SMITH, A. G. 1958. Post-glacial deposits in south Yorkshire and north Lincolnshire. *New Phytol.*, Vol. 57, pp. 19–49.
STRAW, A. 1961a. The erosion surfaces of east Lincolnshire. *Proc. Yorkshire Geol. Soc.*, Vol. 33, pp. 149–172.
— 1961b. Drifts, meltwater channels and ice-margins in the Lincolnshire Wolds. *Trans. Inst. Br. Geogr.*, Publ. No. 29, pp. 115–128.
— 1963. The Quaternary evolution of the lower and middle Trent. *East Midland Geogr.*, Vol. 3, pp. 171–189.
— 1969. Pleistocene events in Lincolnshire: a survey and revised nomenclature. *Trans. Lincolnshire Nat. Union*, Vol. 17, pp. 85–98.
WATTS, W. A. 1959. Pollen spectra from the interglacial deposits at Kirmington, Lincolnshire. *Proc. Yorkshire Geol. Soc.*, Vol. 32, pp. 145–152.

Economic minerals and water supply (Chapter 13)

GRAY, D. A. 1974. Water resources and supply *in* RAYNER, D. H. and HEMINGWAY, J. E. (Editors), *q.v.*
HEMINGWAY, J. E. 1974. Ironstone *in* RAYNER, D. H. and HEMINGWAY, J. E. (Editors), *q.v.*
HULL, J. H. and THOMAS, I. A. 1974. Limestones and dolomites *in* RAYNER, D. H. and HEMINGWAY, J. E. (Editors), *q.v.*
SMITH, D. B. 1974. Evaporites *in* RAYNER, D. H. and HEMINGWAY, J. E. (Editors), *q.v.*
SMITH, E. G. 1974. Constructional materials and miscellaneous mineral products *in* RAYNER, D. H. and HEMINGWAY, J. E. (Editors), *q.v.*

Index

Printed in England for Her Majesty's Stationery Office
by Raithby, Lawrence & Company Ltd. Dd696807 K160